ANCIENT AND MODERN

SHIPS

WOODEN SAILING-SHIPS

BY
SIR GEORGE C. V. HOLMES

PART I

HON. MEMBER I.N.A., WHITWORTH SCHOLAR.
FORMERLY SECRETARY OF THE INSTITUTION OF NAVAL ARCHITECTS

WITH SEVENTY-FOUR ILLUSTRATIONS.

(*Revised.*)

LONDON:
PRINTED FOR HIS MAJESTY'S STATIONERY OFFICE,
By WYMAN AND SONS, Limited, Fetter Lane, E.C.
———
1906.

ISBN-13: 978-1721630288

ISBN-10: 1721630287

PREFACE

An endeavour has been made in this handbook, as far as space and scantiness of material would permit, to trace the history of the development of wooden ships from the earliest times down to our own. Unfortunately, the task has been exceedingly difficult; for the annals of shipbuilding have been very badly kept down to a quite recent period, and the statements made by old writers concerning ships are not only meagre but often extremely inaccurate. Moreover, the drawings and paintings of vessels which have survived from the classical period are few and far between, and were made by artists who thought more of pictorial effect than of accuracy of detail. Fortunately the carvings of the ancient Egyptians were an exception to the above rule. Thanks to their practice of recording and illustrating their history in one of the most imperishable of materials we know more of their ships and maritime expeditions than we do of those of any other people of antiquity. If their draughtsmen were as conscientious in delineating their boats as they were in their drawings of animals and buildings, we may accept the illustrations of Egyptian vessels which have survived into our epoch as being correct in their main features. The researches now being systematically carried out in the Valley of the Nile add, year by year, to our knowledge, andviii already we know enough to enable us to assert that ship building is one of the oldest of human industries, and that there probably existed a sea borne commerce in the Mediterranean long before the building of the Pyramids.

Though the Phœnicians were the principal maritime people of antiquity in the Mediterranean, we know next to nothing of their vessels. The same may be said of the Greeks of the Archaic period. There is, however, ground for hope that, with the progress of research, more may be discovered concerning the earliest types of Greek vessels; for example, during the past year, a vase of about the eighth century b.c. was found, and on it is a representation of a bireme of the Archaic period of quite exceptional interest. As the greater part of this handbook was already in type when the vase was acquired by the British Museum, it has only been possible to reproduce the representation in the Appendix. The drawings of Greek merchant-ships and galleys on sixth and fifth-century vases are merely pictures, which tell us but little that we really want to know. If it had not been for the discovery, this century, that a drain at the Piræus was partly constructed of marble slabs, on which were engraved the inventories of the Athenian dockyards, we should know but little of the Greek triremes of as late a period as the third century b.c. We do not possess a single illustration of a Greek or Roman trireme, excepting only a small one from Trajan's Column, which must not be taken too seriously, as it is obviously pictorial, and was made a century and a half after many-banked ships had gone out of fashion.

In the first eight centuries of our era records and illustrations of ships continue to be extremely meagre. Owing to aix-x comparatively recent discovery we know something of Scandinavian boats. When we consider the way in which the Norsemen overran the seaboard of Europe, it seems probable that their types of vessels were dominant, at any rate in Northern and Western European waters, from the tenth to the twelfth century. From the time of the Norman Conquest down to the reign of Henry VIII. we have to rely, for information about ships, upon occasional notes by the old chroniclers, helped out by a few illustrations taken from ancient corporate seals and from manuscripts. From the time of Henry VIII., onwards, information about warships is much more abundant; but, unfortunately, little is known of the merchant vessels of the Tudor, Stuart, and early Hanoverian periods, and it has not been found possible to trace the origin and development of the various types of merchant sailing-ships now in existence.

The names of the authorities consulted have generally been given in the text, or in footnotes. The author is indebted to Dr. Warre's article on ships, in the last edition of the "Encyclopædia Britannica," and to Mr. Cecil Torr's work, "Ancient Ships," for much information concerning Greek and Roman galleys, and further to "The Royal Navy," a history by Mr. W. Laird Clowes, and the "History of Marine Architecture" by Charnock, for much relating to British warships down to the end of the eighteenth century.

5, Adelphi Terrace, W.C.,
January, 1, 1900.

Note: Original page numbers can be found throughout this text

CONTENTS.

LIST OF ILLUSTRATIONS.

CHAP.　　　.

8

The illustrations marked * are published by kind permission of the Committee of the Egypt Exploration Fund. Those marked † are taken from "The History of Merchant Shipping and Ancient Commerce," and were kindly lent by Messrs. Sampson Low, Marston & Co., Ltd. Those marked ‡ are reproduced from "La Marine Française de 1792 à nos jours," by l'Amiral Paris.

ANCIENT AND MODERN SHIPS.

Part I.

WOODEN SAILING-SHIPS.

CHAPTER I.

INTRODUCTION.

A museum relating to Naval Architecture and Shipbuilding is of the utmost interest to the people of Great Britain, on account of the importance to them of everything that bears on the carrying of their commerce. Every Englishman knows, in a general way, that the commerce of the British Empire is more extensive than that of any other state in the world, and that the British sea-going mercantile marine compares favourably in point of size even with that of all the other countries of the world put together; but few are probably aware of the immense importance to us of these fleets of trading ships, and of the great part which they play in the maintenance of the prosperity of these isles. The shipping industry ranks, after agriculture, as the largest of our national commercial pursuits. There is more capital locked up in it, and more hands are employed in the navigation and construction of ships, their engines and fittings, than in any other trade of the country excepting the tillage of the soil.

The following Table gives the relative figures of the merchant navies of the principal states of the civilised world in the year 1898, and proves at a glance the immense interest to our fellow countrymen of all that affects the technical2 advancement of the various industries connected with shipping: —

Number and Tonnage of Sailing-vessels of over 100 Tons net, and Number and Tonnage of Steamers of over 100 Tons gross, belonging to each of the Countries named, as recorded in Lloyds' Register Book.

Flag.	Total No. of steam and sailing vessels.	Total tonnage of steam (gross) and of sailing vessels (net).
United Kingdom	8,973	12,926,924
Colonies	2,025	1,061,584
Total	10,998	13,988,508
United States of America, including Great Lakes	3,010	2,465,387
Danish	796	511,958
French	1,182	1,242,091
German	1,676	2,453,334
Italian	1,150	875,851
Japanese	841	533,381
Norwegian	2,528	1,694,230
Russian	1,218	643,527
Spanish	701	608,885
Swedish	1,408	605,991
All other countries	2,672	2,050,385
Total	28,180	27,673,528

The part played by technical improvements in the maintenance of our present position cannot be over-estimated; for that position, such as it is, is not due to any inherent permanent advantages possessed by this country. Time was when our mercantile marine was severely threatened by competition from foreign states. To quote the most recent example, about the middle of last century the United States of America fought a well-contested struggle with us for the carrying trade of the world. Shortly after the abolition of the navigation laws, the competition was very severe, and United States ships had obtained almost exclusive possession of the China trade, and of the trade between Europe and North America, and in the year3 1850 the total tonnage of the shipping of the States was 3,535,434, against 4,232,960 tons owned by Great Britain. The extraordinary progress in American mercantile shipbuilding was due, in part, to special circumstances connected with their navigation laws, and in part to the abundance and cheapness of excellent timber; but, even with these advantages, the Americans would never have been able to run such a close race with us for the carrying trade of the world, had it not been for the great technical skill and intelligence of their shipbuilders, who produced vessels which were the envy and admiration of our own constructors. As a proof of this statement, it may be mentioned that, the labour-saving mechanical contrivances adopted by the Americans were such that, on board their famous liners and clippers, twenty men could do the work which in a British ship of equal size required thirty, and, in addition to this advantage, the American vessels could sail faster and carry more cargo in proportion to their registered tonnage than our own vessels. It was not till new life was infused into British naval architecture that we were enabled to conquer the American competition; and then it was only by producing still better examples of the very class of ship which the Americans had been the means of introducing, that we were eventually enabled to wrest from them the China trade. Another triumph in the domain of technical shipbuilding, viz., the introduction and successful development of the iron-screw merchant steamer, eventually secured for the people of this country that dominion of the seas which remains with them to this day.

Among the great means of advancing technical improvements, none takes higher rank than a good educational museum; for it enables the student to learn, as he otherwise cannot learn, the general course which improvements have taken since the earliest times, and hence to appreciate the4 direction which progress will inevitably take in the future. Here he will learn, for instance, how difficulties have been overcome in the past, and will be the better prepared to play his part in overcoming those with which he, in his turn, will be confronted. In such a museum he can study the advantages conferred upon the owner, by the successive changes which have been effected in the materials, construction, and the means of propulsion of ships. He can trace, for instance, the effects of the change from wood to iron, and from iron to steel, in the carrying capacity of ships, and he can note the effects of successive improvements in the propelling machinery in saving weight and space occupied by engines, boilers, and bunkers; and in conferring upon a ship of a given size the power of making longer voyages. Here, too, he can learn how it was that the American clipper supplanted the old English sailing merchantman, and how the screw iron ship, fitted with highly economical engines, has practically driven the clipper from the seas. In fact, with the aid of a good museum the student is enabled to take a bird's-eye view of the whole chain of progress, in which the existing state of things constitutes but a link.

Signs are not wanting that the competition with which British shipowners had to contend in the past will again become active in the near future. The advantages conferred upon us by abundant supplies of iron and by cheap labour will not last for ever. There are many who expect, not without reason, that the abolition or even the diminution of protection in the United States will, when it comes to pass, have the same stimulating effect upon the American shipbuilding industry which the abolition of the old navigation laws had upon our own; and when that day comes Englishmen will find it an advantage to be able to enter the contest equipped with the best attainable technical education and experience.

CHAPTER II.

ANCIENT SHIPS IN THE MEDITERRANEAN AND RED SEAS.

It is not difficult to imagine how mankind first conceived the idea of making use of floating structures to enable him to traverse stretches of water. The trunk of a tree floating down a river may have given him his first notions. He would not be long in discovering that the tree could support more than its own weight without sinking. From the single trunk to a raft, formed of several stems lashed together, the step would not be a long one. Similarly, once it was noticed that a trunk, or log, could carry more than its own weight and float, the idea would naturally soon occur to any one to diminish the inherent weight of the log by hollowing it out and thus increase its carrying capacity; the subsequent improvements of shaping the underwater portion so as to make the elementary boat handy, and to diminish its resistance in the water, and of fitting up the interior so as to give facilities for navigating the vessel and for accommodating in it human beings and goods, would all come by degrees with experience. Even to the present day beautiful specimens exist of such boats, or canoes, admirably formed out of hollowed tree-trunks. They are made by many uncivilized peoples, such as the islanders of the Pacific and some of the tribes of Central Africa. Probably the earliest type of *built-up* boat was made by stretching skins on a frame. To this class belonged the coracle of the Ancient Britons, which is even now in common use on the Atlantic seaboard of Ireland. The transition from a raft to a flat-bottomed boat was a very6 obvious improvement, and such vessels were probably the immediate forerunners of ships.

It is usual to refer to Noah's ark as the oldest ship of which there is any authentic record. Since, however, Egypt has been systematically explored, pictures of vessels have been discovered immensely older than the ark—that is to say, if the date usually assigned to the latter (2840 b.c.) can be accepted as approximately correct; and, as we shall see hereafter (p. 25), there are vessels *now in existence in Egypt which were built* about this very period. The ark was a vessel of such enormous size that the mere fact that it was constructed argues a very advanced knowledge and experience on the part of the contemporaries of Noah. Its dimensions were, according to the biblical version, reckoning the cubit at eighteen inches; length, 450 feet; breadth, 75 feet; and depth, 45 feet. If very full in form its "registered tonnage" would have been nearly 15,000. According to the earlier Babylonian version, the depth was equal to the breadth, but, unfortunately, the figures of the measurements are not legible.

It has been sometimes suggested that the ark was a huge raft with a superstructure, or house, built on it, of the dimensions given above. There does not, however, appear to be the slightest reason for concurring with this suggestion. On the contrary, the biblical account of the structure of the ark is so detailed, that we have no right to suppose that the description of the most important part of it, the supposed raft, to which its power of floating would have been due, would have been omitted. Moreover, the whole account reads like the description of a ship-shaped structure.

Shipbuilding in Egypt.

The earliest information on the building of ships is found, as might be expected, on the Egyptian tombs and monuments.7 It is probable that the valley of the Nile was also the first land bordering on the Mediterranean in which ships, as distinguished from more elementary craft, were constructed. Everything is in favour of such a supposition. In the first place, the country was admirably situated, geographically, for the encouragement of the art of navigation, having seaboards on two important inland seas which commanded the commerce of Europe and Asia. In the next place, the habitable portion of Egypt consisted of a long narrow strip of densely peopled, fertile territory, bordering a great navigable river, which formed a magnificent highway throughout the whole extent of the country. It is impossible to conceive of physical circumstances more conducive to the discovery and development of the arts of building and navigating floating structures. The experience gained on the safe waters of the Nile would be the best preparation for taking the bolder step of venturing on the open seas. The character of the two inland seas which form the northern and eastern frontiers of Egypt was such as to favour, to the greatest extent, the spirit of adventure. As a rule, their waters are relatively calm, and the distances to be traversed to reach other lands are inconsiderable. We know that the ancient Egyptians, at a period which the most modern authorities place at about 7,000 years ago, had already attained to a very remarkable degree of civilisation and to a knowledge of the arts of construction on land which has never since been excelled. What is more natural than to suppose that the genius and science which enabled them to build the Pyramids and their vast temples and palaces, to construct huge works for the regulation of the Nile, and to quarry, work into shape, and move into place blocks of granite weighing in some cases several hundreds of tons, should also lead them to excel in the art of building ships? Not only the physical circumstances, but the habits and the religion of the8 people created a demand, even a necessity, for the existence of navigable floating structures. At the head of the delta of the Nile was the ancient capital, the famous city of Memphis, near to which were built the Pyramids, as tombs in which might be preserved inviolate until the day of resurrection, the embalmed bodies of their kings. The roofs of the burial chambers in the heart of the Pyramids were prevented from falling in, under the great weight of the superincumbent mass, by huge blocks, or beams, of the hardest granite, meeting at an angle above the chambers. The long galleries by which the chambers were approached were closed after the burial by enormous gates, consisting of blocks of granite, which were let down from above, sliding in grooves like the portcullis of a feudal castle. In this way it was hoped to preserve the corpse contained in the chamber absolutely inviolate. The huge blocks of granite, which weighed from 50 to 60 tons each, were supposed to be too heavy ever to be moved again after they had been once lowered into position, and they were so hard that it was believed they could never be pierced. Now, even if we had no other evidence to guide us, the existence of these blocks of granite in the Pyramids would afford the strongest presumption that the Egyptians of that remote time were perfectly familiar with the arts of inland navigation, for the stone was quarried at Assouan, close to the first cataract, 583 miles above Cairo, and could only have been conveyed from the quarry to the building site by water.

In the neighbourhood of Memphis are hundreds of other blocks of granite from Assouan, many of them of enormous size. The Pyramid of Men-kau-Ra, or Mycerinus, built about

14

3633 b.c., was once entirely encased with blocks from Assouan. The Temple of the Sphinx, built at a still earlier date, was formed, to a large extent, of huge pieces of the same material, each measuring $15 \times 5 \times 3 \cdot 2$ feet, and weighing about 189 tons. The mausoleum of the sacred bulls at Sakara contains numbers of Assouan granite sarcophagi, some of which measure $13 \times 8 \times 11$ feet. These are but a few instances, out of the many existing, from which we may infer that, even so far back as the fourth dynasty, the Egyptians made use of the arts of inland navigation. We are, however, fortunately not obliged to rely on inference, for we have direct evidence from the sculptures and records on the ancient tombs. Thanks to these, we now know what the ancient Nile boats were like, and how they were propelled, and what means were adopted for transporting the huge masses of building material which were used in the construction of the temples and monuments.

The art of reading the hieroglyphic inscriptions was first discovered about the year 1820, and the exploration of the tombs and monuments has only been prosecuted systematically during the last five-and-twenty years. Most of the knowledge of ancient Egyptian ships has, therefore, been acquired in quite recent times, and much of it only during the last year or two. This is the reason why, in the old works on shipbuilding, no information is given on this most interesting subject. Knowledge is, however, now being increased every day, and, thanks to the practice of the ancient Egyptians of recording their achievements in sculpture in a material which is imperishable in a dry climate, we possess at the present day, probably, a more accurate knowledge of their ships than we do of those of any other ancient or mediæval people.

By far the oldest boats of which anything is now known were built in Egypt by the people who inhabited that country before the advent of the Pyramid-builders. It is only within the last few years that these tombs have been explored and critically examined. They are now supposed to be of Libyan origin and to date from between 5000 and 6000 b.c. In many of these10 tombs vases of pottery have been discovered, on which are painted rude representations of ships. Some of the latter were of remarkable size and character. Fig. 2 is taken from one of these vases. It is a river scene, showing two boats in procession. The pyramid-shaped mounds in the background represent a row of hills. These boats are evidently of very large size. One of them has 58 oars, or more probably paddles, on each side, and two large cabins amidships, connected by a flying bridge, and with spaces fenced off from the body of the vessel. The steering was, apparently, effected by means of three large paddles on each side, and from the prow of one of the boats hangs a weight, which was probably intended for an anchor. It will be noticed that the two ends of these vessels, like the Nile boats of the Egyptians proper, were not waterborne. A great many representations of these boats have now been discovered. They all have the same leading characteristics, though they differ very much in size. Amongst other peculiarities they invariably have an object at the prow resembling two branches of palm issuing from a stalk, and also a mast carrying an ensign at the after-cabin.

Fig. 2.—The oldest known ships. Between 5000 and 6000 b.c.

Some explorers are of opinion that these illustrations do not represent boats, but fortifications, or stockades of some sort. If we relied only on the rude representations painted on the vases, the question might be a moot one. It has, however, 11 been definitely set at rest by Professor Flinders Petrie, who, in the year 1899, brought back from Egypt very large drawings of the same character, taken, not from vases, but from the tombs themselves. The drawings clearly show that the objects are boats, and that they were apparently very shallow and flat-bottomed. It is considered probable that they were employed in over-sea trade as well as for Nile traffic; for, in the same tombs were found specimens of pottery of foreign manufacture, some of which have been traced to Bosnia.

Fig. 3.—Egyptian boat of the time of the third dynasty.

The most ancient mention of a ship in the world's history is to be found in the name of the eighth king of Egypt after Mena, the founder of the royal race. This king, who was at the head of the second dynasty, was called Betou (Boëthos in Greek), which word signifies the "prow of a ship." Nineteen kings intervened between him and Khufu (Cheops), the builder of the Great Pyramid at Ghizeh. The date of this pyramid is given by various authorities as from about 4235 to 3500 b.c. As the knowledge of Egyptology increases the date is set further and further back, and the late Mariette Pasha, who was one of the greatest authorities on the subject, fixed it at 4235 b.c. About five centuries intervened between the reign of Betou and the date of the Great Pyramid. Hence we can infer that ships were known to the Egyptians of the dynasties sixty-seven centuries ago.

Fortunately, however, we are not obliged to rely on inferences drawn from the name of an individual; we actually12 possess pictures of vessels which, there is every reason to believe, were built before the date of the Great Pyramid.

The boat represented by Fig. 3 is of great interest, as it is by far the oldest specimen of a true Egyptian boat that has yet been discovered. It was copied by the late Mr. Villiers Stuart from the tomb of Ka Khont Khut, situated in the side of a mountain near Kâu-el-Kebîr, on the right bank of the Nile, about 279 miles above Cairo.1 The tomb belongs to a very remote period. From a study of the hieroglyphs, the names of the persons, the forms of the pottery found, and the shape, arrangement, and decoration of the tomb, Mr. Villiers Stuart came to the conclusion that it dates from the third dynasty, and that, consequently, it is older than the Great Pyramid at Ghizeh. If these conclusions are correct, and if Mariette's date for the Great Pyramid be accepted, Fig. 3 represents a Nile boat as used about 6,300 years ago—that is to say, about fifteen centuries before the date commonly accepted for the ark. Mr. Villiers Stuart supposes that it was a dug-out canoe, but from the dimensions of the boat this theory is hardly tenable. It will be noted that there are seven paddlers on each side, in addition to a man using a sounding, or else a punt, pole at the prow, and three men steering with paddles in the stern, while amidships there is a considerable free space, occupied only by the owner, who is armed with a whip, or courbash. The paddlers occupy almost exactly one-half of the total length, and from the space required for each of them the boat must have been quite 56 feet long. It could hardly have been less than seven feet wide, as it contained a central cabin, with sufficient space on either side of the latter for paddlers to sit. If it were a "dug-out," the tree from which it was made must have been brought down the river from tropical 13Africa. There is no reason, however, to suppose anything of the sort; for, if the epoch produced workmen skilful enough to excavate and decorate the tomb, and to carve the statues and make the pottery which it contained, it must also have produced men quite capable of building up a boat from planks.

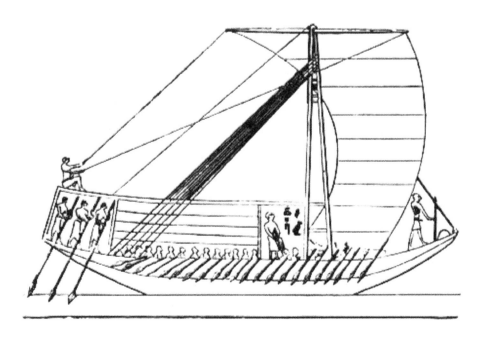

Fig. 4.—Egyptian boat of the time of the fourth dynasty.

The use of sails was also understood at this remote epoch, for it will be noticed that, on the roof of the cabin is lying a mast which has been unshipped. The mast is triangular in shape, consisting of two spars, joined together at the top at an acute angle, and braced together lower down. This form was probably adopted in order to dispense with stays, and thus facilitate shipping and unshipping. It is also worthy of note that this boat appears to have been decked over, as the feet of all those on board are visible above the gunwale. A representation of a very similar boat was found in the tomb of Merâb, a son of Khufu, of the fourth dynasty.

The tombs of Egypt abound in pictures of boats and larger vessels, and many wooden models of them have also been found in the sarcophagi. There is in the Berlin Museum a model of14 a boat similar in general arrangement to the one just described. It is decked over and provided with a cabin amidships, which does not occupy the full width of the vessel. Fig. 4 is a vessel of later date and larger size than that found in the tomb of Ka Khont Khut, but its general characteristics are similar. From the number of paddlers it must have been at least 100 feet in length. In this case we see the mast is erected and a square sail set. The bow and stern also come much higher out of the water. The roof of the cabin is prolonged aft, so as to form a shelter for the steersman and a seat for the man holding the ropes. Similarly it is prolonged forward, so as to provide a shelter for the captain, or owner. The method of steering with oars continued in use for centuries; but in later and larger vessels the steering-oars, which were of great size, were worked by a mechanical arrangement. The illustration was taken originally from a fourth-dynasty tomb at Kôm-el-Ahmars.

There are also extant pictures of Egyptian cattle-boats, formed of two ordinary barges lashed together, with a temporary house, or cattle-shed, constructed across them. The history of Egypt, as inscribed in hieroglyphs on the ancient monuments, relates many instances of huge sarcophagi, statues, and obelisks having been brought down the Nile on ships. The tombs and monuments of the sixth dynasty are particularly rich in such records. In the tomb of Una, who was a high officer under the three kings, Ati, Pepi I., and Mer-en-Ra, are inscriptions which shed a flood of light on Egyptian shipbuilding of this period, and on the uses to which ships were put. In one of them we learn how Una was sent by Pepi to quarry a sarcophagus in a single piece of limestone, in the mountain of Jurra, opposite to Memphis, and to transport it, together with other stones, in one of the king's ships. In another it is related how he headed a military expedition15 against the land of Zerehbah, "to the north of the land of the Hirusha," and how the army was embarked in ships.

In the reign of Pepi's successor, Mer-en-Ra, Una appears to have been charged with the quarrying and transport of the stones destined for the king's pyramid, his sarcophagus, statue, and other purposes. The following passage from the inscriptions on his tomb gives even the number of the ships and rafts which he employed on this work:2—

"His Holiness, the King Mer-en-Ra, sent me to the country of Abhat to bring back a sarcophagus with its cover, also a small pyramid, and a statue of the King Mer-en-Ra, whose pyramid is called Kha-nofer ('the beautiful rising'). And his Holiness sent me to the city of Elephantine to bring back a holy shrine, with its base of hard granite, and the doorposts and cornices of the same granite, and also to bring back the granite posts and thresholds for the temple opposite to the pyramid Kha-nofer, of King Mer-en-Ra. The number of ships destined for the complete transport of all these stones consisted of six broad vessels, three tow-boats, three rafts, and one ship manned with warriors."

Further on, the inscriptions relate how stone for the Pyramid was hewn in the granite quarries at Assouan, and how rafts were constructed, 60 cubits in length and 30 cubits in breadth, to transport the material. The Royal Egyptian cubit was 20·67 inches in length, and the common cubit 18·24 inches. The river had fallen to such an extent that it was not possible to make use of these rafts, and others of a smaller size had to be constructed. For this purpose Una was despatched up the river to the country of Wawa-t, which Brugsch considered to be the modern Korosko. The inscription states—

"His Holiness sent me to cut down four forests in the South, in order to build three large vessels and four towing-vessels out of the acacia wood in the country of Wawa-t. And behold the officials of Araret, Aam, and Mata caused the wood to be cut down for this purpose.

I executed all this in the space of a year. As soon as the waters rose I loaded the rafts with immense pieces of granite for the Pyramid Kha-nofer, of the King Mer-en-Ra."

Mr. Villiers Stuart found several pictures of large ships of this remote period at Kasr-el-Syad on the Nile, about 70 miles below Thebes, in the tomb of Ta-Hotep, who lived in the reigns of Pepi I. and his two successors. These boats were manned with twenty-four rowers, and had two cabins, one amidships and the other astern.3 The same explorer describes the contents of a tomb of the sixth dynasty at Gebel Abû Faida, on the walls of which he observed the painting of a boat with a triple mast (presumably made of three spars arranged like the edges of a triangular pyramid), and a stern projecting beneath the water.

Between the sixth and the eleventh dynasties Egyptian history is almost an utter blank. The monuments contain no records for a period of about 600 years. We are, therefore, in complete ignorance of the progress of shipbuilding during this epoch. It was, however, probably considerable; for, when next the monuments speak it is to give an account of a mercantile expedition on the high seas. In the Valley of Hamâmât, near Coptos, about 420 miles above Cairo, is an inscription on the rocks, dating from the reign of Sankh-ka-Ra, the last king of the eleventh dynasty (about 2800 b.c.), describing an expedition by sea to the famous land of Punt, on the coast of the Red Sea. This expedition is not to be confounded with another, a much more famous one, to the same land, carried out by direction of Queen Hatshepsu of the eighteenth dynasty, about eleven centuries later. Sankh-ka-Ra's enterprise is, however, remarkable as being the first over-sea maritime expedition recorded in the world's history. It may be noted that it took place at about the date usually assigned to Noah's ark.

The town of Coptos was of considerable commercial importance, having been at one end of the great desert route from the Nile to the Red Sea port of Kosseir, whence most of the Egyptian maritime expeditions started. The land of Punt, which was the objective of the expedition, is now considered to be identical with Somaliland. The following extracts from the inscription give an excellent idea of the objects and conduct of the expedition, which was under the leadership of a noble named Hannu, who was himself the author of the inscription:4—

"I was sent to conduct ships to the land of Punt, to fetch for Pharaoh sweet-smelling spices, which the princes of the red land collect out of fear and dread, such as he inspires in all nations. And I started from the City of Coptos, and his Holiness gave the command that the armed men, who were to accompany me, should be from the south country of the Thebaîd."

After describing the arrangements which he made for watering the expedition along the desert route, he goes on to say:—

"Then I arrived at the port Seba, and I had ships of burthen built to bring back products of all kinds. And I offered a great sacrifice of oxen, cows, and goats. And when I returned from Seba I had executed the King's command, for I brought him back all kinds of products which I had met with in the ports of the Holy Land (Punt). And I came back by the road of Uak and Rohan, and brought with me precious stones for the statues of the temples. But such a thing never happened since there were kings; nor was the like of it ever done by any blood relations who were sent to these places since the time (of the reign) of the Sun-god Ra."

From the last sentence of the above quotation we may infer that previous expeditions had been sent to the land of Punt. Communication with this region must, however, have been carried on only at considerable intervals, for we read that 18Hannu had to build the ships required for the voyage. Unfortunately, no representations of these vessels accompany the inscription.

Between the end of the eleventh and the commencement of the eighteenth dynasty, the monuments give us very little information about ships or maritime expeditions. Aahmes, the first king of the latter dynasty, freed Egypt from the domination of the Shepherd Kings by means of a naval expedition on the Nile and the Mediterranean. A short history of this campaign is given in the tomb of another Aahmes, near El Kab, a place on the east bank of the river, 502 miles south of Cairo. This Aahmes was a captain of sailors who served under Sequenen-Ra, King Aahmes, Amenophis I., and Thotmes I. King Aahmes is supposed to have been the Pharaoh of the Old Testament who knew not Joseph. He lived about 1700 b.c.

By far the most interesting naval records of this dynasty are the accounts, in the temple of Dêr-el-Bahari close to Thebes, of the famous expedition to the land of Punt, carried out by order of that remarkable woman Queen Hatshepsu, who was the daughter of Thotmes I., half-sister and wife of Thotmes II., and aunt and step-mother of the famous king Thotmes III. She appears to have been called by her father during his lifetime to share the throne with him, and to have practically usurped the government during the reign of her husband and during the early years of the reign of her nephew.

The expedition to the land of Punt was evidently one of the most remarkable events of her reign. It took place about 1600 b.c. —that is to say, about three centuries before the Exodus. The history of the undertaking is given at great length on the retaining wall of one of the terraces of the temple, and the various scenes and events are illustrated by carvings on the same wall, in as complete a manner as though the expedition19 had taken place in the present time, and had been accompanied by the artists of one of our pictorial newspapers. Fortunately, the great bulk of the carvings and inscriptions remain to this day, and we possess, therefore, a unique record of a trading expedition carried out at this remote period.

The carvings comprise representations of the ships going out. The landing at the "incense terraced-mountain," and the meeting with the princes and people of this strange land, are also shown. We have pictures of their pile dwellings, and of the trees and animals of the country, and also portraits of the King of Punt, of his wife and children. Lastly, we have representations of the ships returning to Egypt, laden with the precious incense of the land and with other merchandise, and also of the triumphant reception of the members of the expedition at Thebes.

One of the inscriptions relates as follows:5—

"The ships were laden to the uttermost with the wonderful products of the land of Punt, and with the different precious woods of the divine land, and with heaps of the resin of incense, with fresh incense trees, with ebony, (objects) of ivory set in pure gold from the land of the 'Amu, with sweet woods, Khesit-wood, with Ahem incense, holy resin, and paint for the eyes, with dog-headed apes, with long-tailed monkeys and greyhounds, with leopard-skins, and with natives of the country, together with their children. Never was the like brought to any king (of Egypt) since the world stands."

The boast contained in the concluding sentence was obviously not justified, as we know the same claims were made in the inscription in the valley of Hammamât, describing the previous expedition to Punt, which took place eleven centuries earlier.

From the frontispiece, Fig. 1, we can form an accurate idea of the ships used in the Red Sea trade in the time of the 20eighteenth dynasty. They were propelled by rowers instead of by paddlers, as in all the previous examples. There were fifteen rowers on each side, and, allowing four feet for the distance between each seat, and taking account of the length of the overhanging portions at bow and stern, the length of each vessel could have been little short of a hundred feet. They were apparently decked over and provided with raised cabins at the two extremities. The projections marked along the sides may indicate the ends of beams, or they may, as some writers have supposed, have been pieces of timber against which the oars could be worked in narrow and shallow water.

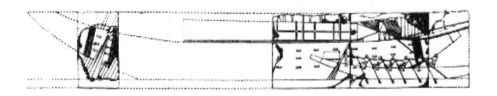

Fig. 5.—Nile barge carrying obelisks. About 1600 b.c.

These vessels were each rigged with a huge square sail. The spars carrying the sail were as long as the boats themselves, and were each formed of two pieces spliced together in the middle. The stems and sterns were not waterborne. In order to prevent the vessel from hogging under the influence of the weights of the unsupported ends, a truss was employed, similar in principle and object to those used to this day in American river steamers. The truss was formed by erecting four or more pillars in the body of the vessel, terminating at a height21 of about six feet above the gunwale, in crutches. A strong rope running fore and aft was passed over these crutches and also round the mast, the two ends of the rope having been so arranged as to gird and support the stem and stern respectively.

The Temple of Dêr-el-Bahari contained also a most interesting illustrated account of the transport of two great obelisks down the Nile in the reign of the same queen. Unfortunately, parts of the description and of the carvings have been lost, but enough remains to give us a very clear idea of the vessels employed and of the method of transport. Fig. 5 shows the type of barge employed to carry the obelisks, of which there were two. The dotted lines show the portions of the carving which are at present missing. The restoration was effected by Monsieur Edouard Naville.6 The restoration is by no means conjectural. The key to it was furnished by a hieroglyph in the form of the barge with the obelisks on deck. Some of these obelisks were of very large size. There are two, which were hewn out of granite for Queen Hatshepsu, still at the Temple of Karnak. They may, very possibly, be the two which are referred to in the description at Dêr-el-Bahari. One of them is 98 feet and the other 105 feet in height. The larger of the two has been calculated to weigh 374 tons, and the two together may have weighed over 700 tons. To transport such heavy stones very large barges would have been required. Unfortunately, the greater portion of the inscription describing the building of these boats has been lost, but what remains states that orders were given to collect "sycamores from the whole land (to do the) work of building a very great boat." There is, however, an inscription still intact in the tomb of an ancient Egyptian named Anna, who lived in the reigns of the three kings Thotmes (and therefore also during 22that of Queen Hatshepsu), which relates that, having to transport two obelisks for Thotmes I., he built a boat 120 cubits long and 40 cubits wide. If the royal cubit of 20·72 inches was referred to, the dimensions of the boat would have been 200 feet long by 69 feet wide. This is possibly the very boat illustrated on the walls of Dêr-el-Bahari; for, it having evidently been a matter of some difficulty to collect the timber necessary to build so large a vessel, it seems only natural to suppose that it would be carefully preserved for the future transport of similar obelisks. If, however, it was found necessary to construct a new boat in order to transport Queen Hatshepsu's obelisks, we may be fairly certain that it was larger than the one whose dimensions are given above, for the taller of her two obelisks at Karnak is the largest that has been found in Egypt in modern times. The obelisk of rose granite of Thotmes I., still at Karnak, is 35 feet shorter, being 70 feet, or exactly the same height as the one called Cleopatra's Needle, now on the Thames Embankment.

The barge shown in Fig. 5 was strengthened, apparently, with three tiers of beams; it was steered by two pairs of huge steering-oars, and was towed by three parallel groups, each consisting of ten large boats. There were 32 oarsmen to each boat in the two wing groups, and 36 in each of the central groups: there were, therefore, exactly one thousand oars used in all. The towing-cable started from the masthead of the foremost boat of each group, and thence passed to the bow of the second one, and so on, the stern of each boat being left perfectly free, for the purpose, no doubt, of facilitating the steering. The flotilla was accompanied by five smaller boats, some of which were used by the priests, while the others were despatch vessels, probably used to keep up communications with the groups of tugs.

There are no other inscriptions, or carvings, that have as yet23 been discovered in Egypt which give us so much information regarding Egyptian ships as those on the Temple at Dêr-el-Bahari. From time to time we read of naval and mercantile expeditions, but illustrations of the ships and details of the voyages are, as a rule, wanting. We know that Seti I., of the nineteenth dynasty, whose reign commenced about 1366 b.c., was a great encourager of commerce. He felled timber in Lebanon for building ships, and is said to have excavated a canal between the Nile and the Red Sea. His successor, the famous Ramses II., carried on wars by sea, as is proved by the inscriptions in the Temple at Abû Simbel in Nubia, 762 miles above Cairo.

In the records of the reign of Ramses III., 1200 b.c., we again come upon illustrations of ships in the Temple of Victory at Medînet Habû, West Thebes. The inscriptions describe a great naval victory which this king won at Migdol, near the Pelusiac mouth of the Nile, over northern invaders, probably Colchians and Carians. Fig. 6 shows one of the battleships. It is probably more a symbolical than an exact representation, nevertheless it gives us some valuable information. For instance, we see that the rowers were protected against the missiles of their adversaries by strong bulwarks, and the captain occupied a crow's nest at the masthead.

Ramses III. did a great deal to develop Egyptian commerce. His naval activities were by no means confined to the Mediterranean, for we read that he built a fleet at Suez, and traded with the land of Punt and the shores of the Indian Ocean. Herodotus states that, in his day, the docks still existed at the head of the Arabian Gulf where this Red Sea fleet was built.

Pharaoh Nekau (Necho), who reigned from 612 to 596 b.c., and who defeated Josiah, King of Judah, was one of the kings of Egypt who did most to encourage commerce. He commenced a canal to join the Pelusiac branch of the Nile at24 Bubastis with the Red Sea, but never finished it. It was under his directions that the Phœnicians, according to Herodotus, made the voyage round Africa referred to on p. 27. When Nekau abandoned the construction of the canal he built two fleets of triremes, one for use in the Mediterranean, and the other for the Red Sea. The latter fleet was built in the Arabian Gulf.

Fig. 6.—Battleship of Ramses III. About 1200 b.c.

In later times the seaborne commerce of Egypt fell, to a large extent, into the hands of the Phœnicians and Greeks.

Herodotus (484 to 423 b.c.) gives an interesting account of the Nile boats of his day, and of the method of navigation of the river.7

"Their boats, with which they carry cargoes, are made of the thorny acacia.... From this tree they cut pieces of wood about two cubits in length, and arrange them like bricks, fastening the boat together by a great number of long bolts through the two-cubit pieces; and when they have thus fastened the boat together they lay cross-pieces over the top, using no ribs for the sides; and within they caulk the seams with papyrus. They make one steering-oar for it, which is passed through the bottom of the boat, and they have a mast of acacia and sails of papyrus. These boats cannot sail up the river unless 25there be a very fresh wind blowing, but are towed.... Down stream they travel as follows: they have a door-shaped crate, made of tamarisk wood and reed mats sewn together, and also a stone of about two talents' weight, bored with a hole; and of these the boatman lets the crate float on in front of the boat, fastened with a rope, and the stone drag behind by another rope. The crate then, as the force of the stream presses upon it, goes on swiftly and draws on the ... boats, ... while the stone, dragging after it behind and sunk deep in the water, keeps its course straight."

In connection with this account it is curious to note that, at so late a period as the time of Herodotus, papyrus was used for the sails of Nile boats, for we know that, for many centuries previously, the Egyptians were adepts in the manufacture of linen, and actually exported fine linen to Cyprus to be used as sail-cloth.

Before concluding this account of shipbuilding in ancient Egypt, it may be mentioned that, in the year 1894, the French Egyptologist, Monsieur J. de Morgan, discovered several Nile boats of the time of the twelfth dynasty (2850 b.c.) admirably preserved in brick vaults at Dashûr, a little above Cairo, on the left bank of the river. The site of these vaults is about one hour's ride from the river and between 70 and 80 feet above the plain. The boats are about 33 feet long, 7 to 8 feet wide, and 2½ to 3 feet deep. As there were neither rowlocks nor masts, and as they were found in close proximity to some Royal tombs, it is considered probable that they were funeral boats, used for carrying royal mummies across the river. They are constructed of planks of acacia and sycamore, about three inches thick, which are dovetailed together and fastened with trenails. There are floors, but no ribs. In this respect the account of Herodotus is remarkably confirmed. The method of construction was so satisfactory that, although they are nearly 5,000 years old, they held rigidly together after their supports had been removed by Monsieur de Morgan. They were steered by two large paddles. The discovery of these 26 boats is of extraordinary interest, for they were built at the period usually assigned to Noah's ark. It is a curious fact that they should have been found so far from the river, but we know from other sources—such as the paintings found in papyrus books—that it was the custom of the people to transport the mummies of royal personages, together with the funeral boats, on sledges to the tomb.

The famous galleys of the Egypt of the Ptolemies belonged to the period of Greek and Roman naval architecture, and will be referred to later.

From the time of the ancient Egyptian vessels there is no record whatever of the progress of naval architecture till we come to the period of the Greeks, and even the early records relating to this country are meagre in the extreme. The Phœnicians were among the first of the races who dwelt on the Mediterranean seaboard to cultivate a seaborne commerce, and to them, after the Egyptians, is undoubtedly due the early progress made in sea-going ships. This remarkable people is said to have originally come to the Levant from the shores of the Persian Gulf. They occupied a strip of territory on the seaboard to the north of Palestine, about 250 miles long and of the average width of only 12 miles. The chief cities were Tyre and Sidon. There are only three representations known to be in existence of the Phœnician ships. They must have been of considerable size, and have been well manned and equipped, for the Phœnicians traded with every part of the then known world, and founded colonies—the principal of which was Carthage—at many places along the coast-line of the Mediterranean. A proof of the size and seaworthiness of their ships was the fact that they made very distant voyages across notoriously stormy seas; for instance, to Cornwall in search of tin, and probably also to the south coast of Ireland. They also coasted along the western shores of Africa. Somewhere 27 between the years 610 and 594 b.c. some Phœnician ships, acting under instructions from Pharaoh Nekau, are said to have circumnavigated Africa, having proceeded from the Indian to the Southern Ocean, and thence round by the Atlantic and through the Pillars of Hercules home. The voyage occupied more than two years, a circumstance which was due to the fact that they always landed in the autumn and sowed a tract of country with corn, and waited on shore till it was fit to cut. In the time of Solomon the joint fleets of the Israelites and Phœnicians made voyages from the head of the Red Sea down the coasts of Arabia and Eastern Africa, and even to Persia and Beluchistan, and probably also to India. The Phœnicians were not only great traders themselves, but they manned the fleets of other nations, and built ships for

25

other peoples, notably for the Egyptians and Persians. It is unfortunate that we have so few representations of the Phœnician ships, but we are justified in concluding that they were of the same general type as those which were used by the Greeks, the Carthaginians, and eventually by the Romans. The representations of their vessels known to be in existence were found28 by the late Sir Austin Layard in the palace built by King Sennacherib at Kouyunjik, near Nineveh, about 700 b.c. One of these is shown in Fig. 7. Though they were obviously rather symbols of ships than faithful representations, we can, nevertheless, gather from them that the warship was a galley provided with a ram, and fitted with a mast carrying a single square sail; there were also two banks of oars on each side. The steering was accomplished by two large oars at the stern, and the fighting troops were carried on a deck or platform raised on pillars above the heads of the rowers.

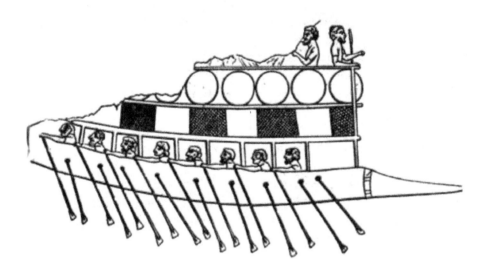

Fig. 7.—Portion of a Phœnician galley. About 700 b.c. *From Kouyunjik (Nineveh).*

Shipbuilding in Ancient Greece and Rome.

In considering the history of the development of shipbuilding, we cannot fail to be struck with the favourable natural conditions which existed in Greece for the improvement of the art. On the east and west the mainland was bordered by inland seas, studded with islands abounding in harbours. Away to the north-east were other enclosed seas, which tempted the enterprise of the early navigators. One of the cities of Greece proper, Corinth, occupied an absolutely unique position for trade and colonization, situated as it was on a narrow isthmus commanding two seas. The long narrow Gulf of Corinth opening into the Mediterranean, and giving access to the Ionian Islands, must have been a veritable nursery of the art of navigation, for here the early traders could sail for long distances, in easy conditions, without losing sight of land. The Gulf of Ægina and the waters of the Archipelago were equally favourable. The instincts of the people were commercial, and their necessities made them colonizers on a vast scale; moreover, they had at their disposal the experience in the arts of navigation, acquired from time immemorial, by the Egyptians and Phœnicians. Nevertheless, with all these circumstances in their favour, the Greeks,29 at any rate up to the fourth century b.c., appear to have contributed nothing to the improvement of shipbuilding.8 The Egyptians and Phœnicians both built triremes as early as 600 b.c., but this class of vessel was quite the exception in the Greek fleets which fought at Salamis 120 years later.

The earliest naval expedition mentioned in Greek history is that of the allied fleets which transported the armies of Hellas to the siege of Troy about the year 1237 b.c. According to the Greek historians, the vessels used were open boats, decks not having been introduced into Greek vessels till a much later period.

The earliest Greek naval battle of which we have any record took place about the year 709 b.c., over 500 years after the expedition to Troy and 1,000 years after the battle depicted in the Temple of Victory at Thebes. It was fought between the Corinthians and their rebellious colonists of Corcyra, now called Corfu.

Some of the naval expeditions recorded in Greek history were conceived on a gigantic scale. The joint fleets of Persia and Phœnicia which attacked and conquered the Greek colonies in Ionia consisted of 600 vessels. This expedition took place in the year 496 b.c. Shortly afterwards the Persian commander-in-chief, Mardonius, collected a much larger fleet for the invasion of Greece itself.

After the death of Cambyses, his successor Xerxes collected a fleet which is stated to have numbered 4,200 vessels, of which 1,200 were triremes. The remainder appears to have been divided into two classes, of which the larger were propelled with twenty-five and the smaller with fifteen oars a-side. This fleet, after many misfortunes at sea, and after gaining a hard-fought victory over the Athenians, was finally destroyed by 30the united Greek fleet at the ever-famous battle of Salamis. The size of the Persian monarch's fleet was in itself a sufficient proof of the extent of the naval power of the Levantine states; but an equally convincing proof of the maritime power of another Mediterranean state, viz., Carthage, at that early period—about 470 b.c. —is forthcoming. This State equipped a large fleet, consisting of 3,000 ships, against the Greek colonies in Sicily; of these 2,000 were fighting galleys, and the remainder transports on which no less than 300,000 men were embarked. This mighty armada was partly destroyed in a great storm. All the transports were wrecked, and the galleys were attacked and totally destroyed by the fleets of the Greek colonists under Gelon on the very day, according to tradition, on which the Persians were defeated at Salamis. Out of the entire expedition only a few persons returned to Carthage to tell the tale of their disasters.

The foregoing account will serve to give a fair idea of the extent to which shipbuilding was carried on in the Mediterranean in the fifth century before the Christian era.

We have very little knowledge of the nature of Greek vessels previously to 500 b.c. 9 Thucydides says that the ships engaged on the Trojan expedition were without decks.

According to Homer, 1,200 ships were employed, those of the Bœotians having 120 men each, and those of Philoctetes 50 men each. Thucydides also relates that the earliest Hellenic triremes were built at Corinth, and that Ameinocles, a Corinthian naval architect, built four ships for the Samians about 700 b.c.; but triremes did not become common until the time of the Persian War, except in Sicily and Corcyra (Corfu), in which states considerable numbers were in use a little time before the war broke out.

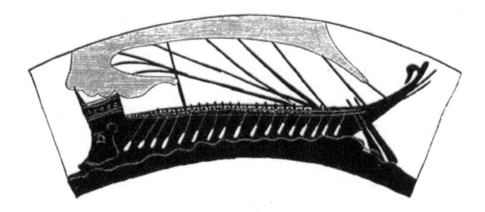

Fig. 8.—Greek unireme. About 500 b.c.

Fig. 8 is an illustration of a single-banked Greek galley of the date about 500 b.c., taken from an Athenian painted vase now in the British Museum. The vessel was armed with a ram; seventeen oars a-side are shown. There is no space on the vase to show in detail the whole of the mast and rigging, but their presence is indicated by lines.

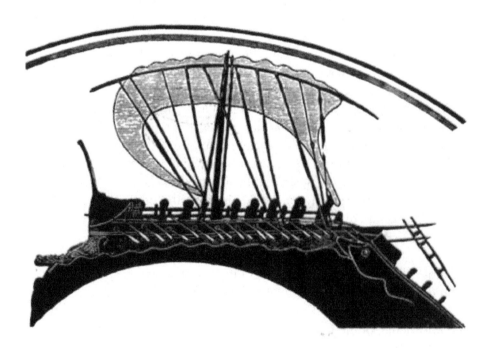

Fig. 9.—Greek bireme. About 500 b.c.

Fig. 9 is a representation of a Greek bireme of about the date 500 b.c. —that is to say, of the period immediately preceding the Persian War. It is taken from a Greek vase in the32 British Museum, which was found at Vulci in Etruria. It is one of the very few representations now in existence of ancient Greek biremes. It gives us far less information than we could wish to have. The vessel has two banks of oars, those of the upper tier passing over the gunwale, and those of the lower passing through oar-ports. Twenty oars are shown by the artist on each side, but this is probably not the exact number used. Unfortunately the rowers of the lower tier are not shown in position. The steering was effected by means of two large oars at the stern, after the manner of those in use in the Egyptian ships previously described. This is proved by another illustration of a bireme on the same vase, in which the steering oars are clearly seen. The vessel had a strongly marked forecastle and a ram fashioned in the shape of a boar's head. It is a curious fact that Herodotus, in his history (Book III.), mentions that the Samian ships carried beaks, formed to resemble the head of a wild boar, and he relates how the Æginetans beat some Samian colonists in a sea-fight off Crete, and sawed off the boar-head beaks from the captured33 galleys, and deposited them in a temple in Ægina. This sea-fight took place about the same time that the vases were manufactured, from which Figs. 8 and 9 are copied. There was a single mast with a very large yard carrying a square sail. The stays are not shown, but Homer says that the masts of early Greek vessels were stayed fore and aft.

Fig. 10.—Fragment of a Greek galley showing absence of deck. About 550 b.c.

It is impossible to say whether this vessel was decked. According to Thucydides, the ships which the Athenians built at the instigation of Themistocles, and which they used at Salamis, were not fully decked. That Greek galleys were sometimes without decks is proved by Fig. 10, which is a copy of a fragment of a painting of a Greek galley on an Athenian vase now in the British Museum, of the date of about 550 b.c. It is perfectly obvious, from the human figures in the galley, that there was no deck. Not even the forecastle was covered in. The galleys of Figs. 8 and 9 had, unlike the Phœnician bireme of Fig. 7, no fighting-deck for the use of the soldiers. There was also no protection for the upper-tier rowers, and in this respect they were inferior to the Egyptian ship shown in Fig. 6. It is probable that Athenian ships at Salamis also had no fighting, or flying decks for the

29

use of the soldiers; for, according to Thucydides, Gylippos, when exhorting the Syracusans, nearly sixty years later, in 413 b.c., said, "But to them (the Athenians) the employment of troops on deck is a novelty." Against this view, however, it must be stated that there are now in existence at Rome two grotesque pictures of Greek galleys on a painted vase, dating from about 550 b.c., in which the soldiers are clearly depicted standing and fighting upon a flying deck. Moreover, Thucydides, in describing a sea-fight between the Corinthians and the Corcyreans in 432 b.c., mentions that the decks of both fleets were crowded with heavy infantry archers and javelin-men, "for their naval engagements were still of the old clumsy sort." Possibly this34 last sentence gives us a clue to the explanation of the apparent discrepancy. The Athenians were, as we know, expert tacticians at sea, and adopted the method of ramming hostile ships, instead of lying alongside and leaving the fighting to the troops on board. They may, however, have been forced to revert to the latter method, in order to provide for cases where ramming could not be used; as, for instance, in narrow harbours crowded with shipping, like that of Syracuse.

It is perfectly certain that the Phœnician ships which formed the most important part of the Persian fleet at Salamis carried fighting-decks. We have seen already (p. 28) that they used such decks in the time of Sennacherib, and we have the distinct authority of Herodotus for the statement that they were also employed in the Persian War; for, he relates that Xerxes returned to Asia in a Phœnician ship, and that great danger arose during a storm, the vessel having been top-heavy owing to the deck being crowded with Persian nobles who returned with the king.

Fig. 11.—Galley showing deck and superstructure. About 600 b.c. *From an Etruscan imitation of a Greek vase.*

Fig. 11, which represents a bireme, taken from an ancient Etruscan imitation of a Greek vase of about 600 b.c., clearly shows soldiers fighting, both on the deck proper and on a raised, or flying, forecastle.

In addition to the triremes, of which not a single illustration35 of earlier date than the Christian era is known to be in existence, both Greeks and Persians, during the wars in the early part of the fifth century b.c., used fifty-oared ships called penteconters, in which the oars were supposed to have been arranged in one tier. About a century and a half after the battle of Salamis, in 330 b.c., the Athenians commenced to build ships with four banks, and five years later they advanced to five banks. This is proved by the extant inventories of

30

the Athenian dockyards. According to Diodoros, they were in use in the Syracusan fleet in 398 b.c. Diodoros, however, died nearly 350 years after this epoch, and his account must, therefore, be received with caution.

The evidence in favour of the existence of galleys having more than five superimposed banks of oars is very slight.

Alexander the Great is said by most of his biographers to have used ships with five banks of oars; but Quintus Curtius states that, in 323 b.c., the Macedonian king built a fleet of seven-banked galleys on the Euphrates. Quintus Curtius is supposed by the best authorities to have lived five centuries after the time of Alexander, and therefore his account of these ships cannot be accepted without question.

It is also related by Diodoros that there were ships of six and seven banks in the fleet of Demetrios Poliorcetes at a battle off Cyprus in 306 b.c., and that Antigonos, the father of Poliorcetes, had ships of eleven and twelve banks. We have seen, however, that Diodoros died about two and a half centuries after this period. Pliny, who lived from 61 to 115 a.d., increases the number of banks in the ships of the opposing fleets at this battle to twelve and fifteen banks respectively. It is impossible to place any confidence in such statements.

Theophrastus, a botanist who died about 288 b.c., and who was therefore a contemporary of Demetrios, mentions in his36 history of plants that the king built an eleven-banked ship in Cyprus. This is one of the very few contemporary records we possess of the construction of such ships. The question, however, arises, Can a botanist be accepted as an accurate witness in matters relating to shipbuilding? The further question presents itself, What meaning is intended to be conveyed by the terms which we translate as ships of many banks? This question will be reverted to hereafter.

In one other instance a writer cites a document in which one of these many-banked ships is mentioned as having been in existence during his lifetime. The author in question was Polybios, one of the most painstaking and accurate of the ancient historians, who was born between 214 and 204 b.c., and who quotes a treaty between Rome and Macedon concluded in 197 b.c., in which a Macedonian ship of sixteen banks is once mentioned. This ship was brought to the Tiber thirty years later, according to Plutarch and Pliny, who are supposed to have copied a lost account by Polybios. Both Plutarch and Pliny were born more than two centuries after this event. If the alleged account by Polybios had been preserved, it would have been unimpeachable authority on the subject of this vessel, as this writer, who was, about the period in question, an exile in Italy, was tutor in the family of Æmilius Paulus, the Roman general who brought the ship to the Tiber.

The Romans first became a naval power in their wars with the Carthaginians, when the command of the sea became a necessity of their existence. This was about 256 b.c. At that time they knew nothing whatever of shipbuilding, and their early war-vessels were merely copies of those used by the Carthaginians, and these latter were no doubt of the same general type as the Greek galleys. The first Roman fleet appears to have consisted of quinqueremes.

The third century b.c. is said to have been an era of gigantic37 ships. Ptolemy Philadelphos and Ptolemy Philopater, who reigned over Egypt during the greater part of that century, are alleged to have built a number of galleys ranging from thirteen up to forty banks. The evidence in this case is derived from two unsatisfactory sources. Athenæos and Plutarch quote one Callixenos of Rhodes, and Pliny quotes one Philostephanos of Cyrene, but very little is known about either Callixenos or Philostephanos. Fortunately, however, Callixenos gives details about the size of the forty-banker, the length of her longest oars, and the number of her crew, which enables us to gauge his value as an authority, and to pronounce his story to be incredible (see p. 45).

Whatever the arrangement of their oars may have been, these many-banked ships appear to have been large and unmanageable, and they finally went out of fashion in the year 31 b.c., when Augustus defeated the combined fleets of Antony and Cleopatra at the battle of Actium. The vessels which composed the latter fleets were of the many-banked order, while Augustus had adopted the swift, low, and handy galleys of the Liburni, who were a seafaring and piratical people from Illyria on the Adriatic coast. Their vessels were originally single-bankers, but afterwards it is said that two banks were adopted. This statement is borne out by the evidence of Trajan's Column, all the galleys represented on it, with the exception of one, being biremes.

Augustus gained the victory at Actium largely owing to the handiness of his Liburnian galleys, and, in consequence, this type was henceforward adopted for Roman warships, and ships of many banks were no longer built. The very word "trireme" came to signify a warship, without reference to the number of banks of oars.

After the Romans had completed the conquest of the38 nations bordering on the Mediterranean, naval war ceased for a time, and the fighting navy of Rome declined in importance. It was not till the establishment of the Vandal kingdom in Africa under Genseric that a revival in naval warfare on a large scale took place. No changes in the system of marine architecture are recorded during all these ages. The galley, considerably modified in later times, continued to be the principal type of warship in the Mediterranean till about the sixteenth century of our era.

Ancient Merchant-ships.

Little accurate information as we possess about the warships of the ancients, we know still less of their merchant-vessels and transports. They were unquestionably much broader, relatively, and fuller than the galleys; for, whereas the length of the latter class was often eight to ten times the beam, the merchant-ships were rarely longer than three or four times their beam. Nothing is known of the nature of Phœnician merchant-vessels. Fig. 12 is an illustration of an Athenian merchant-ship of about 500 b.c. It is taken from the same painted vase as the galley shown on Fig. 9. If the illustration can be relied on, it shows that these early Greek sailing-ships were not only relatively short, but very deep. The forefoot and dead wood aft appear to have been cut away to an extraordinary extent, probably for the purpose of increasing the handiness. The rigging was of the type which was practically universal in ancient ships.

Fig. 13 gives the sheer draught or side elevation, the plan, elevations of the bow and stem, and a midship section of a Roman vessel, which from her proportions and the shape of bow is supposed to have been a merchant-ship. The illustration is taken from a model presented to Greenwich Hospital by Lord Anson. The original model was of white marble, and39 was found in the Villa Mattei in Rome, in the sixteenth century.

We know from St. Paul's experiences, as described in the Acts of the Apostles, that Mediterranean merchant-ships must often have been of considerable size, and that they were capable of going through very stormy voyages. St. Paul's ship contained a grain cargo, and carried 276 human beings.

Fig. 12.—Greek merchant-ship. About 500 b.c.

In the merchant-ships oars were only used as an auxiliary means of propulsion, the principal reliance being placed on masts and sails. Vessels of widely different sizes were in use, the larger carrying 10,000 talents, or 250 tons of cargo. Sometimes, however, much bigger ships were used. For instance, Pliny mentions a vessel in which the Vatican obelisk and its pedestal, weighing together nearly 500 tons, were brought from Egypt to Italy about the year 50 a.d. It is further stated that this vessel carried an additional cargo of 800 tons of lentils to keep the obelisk from shifting on board.

Lucian, writing in the latter half of the second century a.d.,40 mentions, in one of his Dialogues, the dimensions of a ship which carried corn from Egypt to the Piræus. The figures are: length, 180 ft.; breadth, nearly 50 ft.; depth from deck to bottom of hold, 43½ ft. The latter figure appears to be incredible. The other dimensions are approximately those of the *Royal George*, described on p. 126.

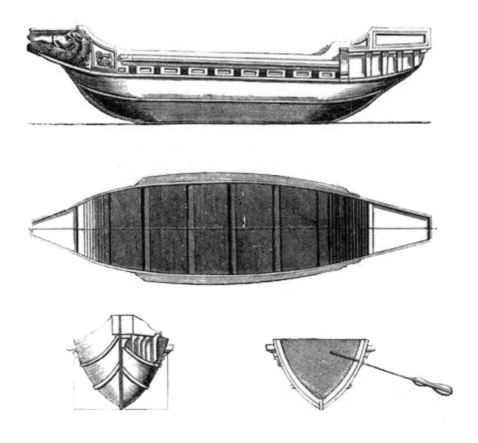

Fig. 13.—Roman merchant-ship.

Details of the Construction of Greek and Roman Galleys.

It is only during the present century that we have learned, with any certainty, what the ancient Greek galleys were like. In the year 1834 a.d. it was discovered that a drain at the Piræus had been constructed with a number of slabs bearing inscriptions, which, on examination, turned out to be the 41 inventories of the ancient dockyard of the Piræus. From these inscriptions an account of the Attic triremes has been derived by the German writers Boeckh and Graser. The galleys all appear to have been constructed on much the same model, with interchangeable parts. The dates of the slabs range from 373 to 323 b.c., and the following description must be taken as applying only to galleys built within this period.

The length, exclusive of the beak, or ram, must have been at least 126 ft., the ram having an additional length of 10 ft. The length was, of course, dictated by the maximum number of oars in any one tier, by the space which it was found necessary to leave between each oar, and by the free spaces between the foremost oar and the stem, and the aftermost oar and the stern of the ship. Now, as it appears further on, the maximum number of oars in any tier in a trireme was 62 in the top bank, which gives 31 a side. If we allow only 3 ft. between the oars we must allot at least 90 ft. to the portion of the vessel occupied by the rowers. The free spaces at stem and stern were, according to the representations of those vessels which have come down to us, about 7/24th of the whole; and, if we accept this

34

proportion, the length of a trireme, independently of its beak, would be about 126 ft. 6 in. If the space allotted to each rower be increased, as it may very reasonably be, the total length of the ship would also have to be increased proportionately. Hence it is not surprising that some authorities put the length at over 140 ft. It may be mentioned in corroboration, that the ruins of the Athenian docks at Zea show that they were originally at least 150 ft. long. They were also 19 ft. 5 in. wide. The breadth of a trireme at the water-line, amidships, was about 14 ft., perhaps increasing somewhat higher up, the sides tumbled home above the greatest width. These figures give the width of the hull proper, exclusive of an outrigged gangway, or42 deck, which, as subsequently explained, was constructed along the sides as a passage for the soldiers and seamen. The draught was from 7 to 8 ft.

Such a vessel carried a crew of from 200 to 225, of whom 174 were rowers, 20 seamen to work the sails, anchors, etc., and the remainder soldiers. Of the rowers, 62 occupied the upper, 58 the middle and 54 the lower tier. Many writers have supposed that each oar was worked by several rowers, as in the galleys of the Middle Ages. This, however, was not the case, for it has been conclusively proved that, in the Greek galleys, up to the class of triremes, at any rate, there was only one man to each oar. For instance, Thucydides, describing the surprise attack intended to be delivered on the Piræus, and actually delivered against the island of Salamis by the Peloponnesians in 429 b.c., relates that the sailors were marched from Corinth to Nisæa, the harbour of Megara, on the Athenian side of the isthmus, in order to launch forty ships which happened to be lying in the docks there, and that *each* sailor carried his cushion and his oar, with its thong, on his march. We have, moreover, a direct proof of the size of the longest oars used in triremes, for the inventories of the Athenian dockyards expressly state that they were 9½ cubits, or 13 ft. 6 in. in length. The reason why the oars were arranged in tiers, or banks, one above the other was, no doubt, that, in this way, the propelling power could be increased without a corresponding increase in the length of the ships. To make a long sea-going vessel sufficiently strong without a closed upper deck would have severely taxed the skill of the early shipbuilders. Moreover, long vessels would have been very difficult to manœuvre, and in the Greek mode of fighting, ramming being one of the chief modes of offence, facility in manœuvring was of prime importance. The rowers on each side sat in the same vertical longitudinal plane, and consequently the length of the inboard43 portions of the oars varied according as the curve of the vessel's side approached or receded from this vertical plane. The seats occupied by the rowers in the successive tiers were arranged one above the other in oblique lines sloping upwards towards the stem, as shown in Figs. 14 and 15. The vertical distance between the seats was about 2 ft. The horizontal gap between the benches in each tier was about 3 ft. The seats were some 9 in. wide, and foot-supports were fixed to each for the use of the rower next above and behind. The oars were so arranged that the blades in each tier all struck the water in the same fore and aft line. The lower oar-ports were about 3 ft., the middle 4¼ ft., and the upper 5½ ft., above the water. The water was prevented from entering the ports by means of leather bags fastened round the oars and to the sides of the oar-ports. The upper oars were about 14 ft. long, the middle 10 ft., and the lower 7½ ft., and in addition to these there were a few extra oars which were occasionally worked from the platform, or deck, above the upper tier, probably by the seamen and soldiers when they were not otherwise occupied. The benches for the rowers extended from the sides to timber supports, inboard, arranged in vertical planes fore and aft. There were two sets of these timbers, one belonging to each side of the ship, and separated by a space of 7 ft. These timbers also connected the upper and lower decks together. The latter was about 1 ft. above the water-line. Below the lower deck was the hold which contained the ballast, and in which the apparatus for baling was fixed.

In addition to oars, sails were used as a means of propulsion whenever the wind was favourable, but not in action.

The Athenian galleys had, at first, one mast, but afterwards, it is thought, two were used. The mainmast was furnished with a yard and square sail.

The upper deck, which was the fighting-platform previously44 mentioned, was originally a flying structure, and, perhaps, did not occupy the full width of the vessel amidships. At the bow, however, it was connected by planking with the sides of the ship, so as to form a closed-in space, or forecastle. This forecastle would doubtless have proved of great use in keeping the ship dry during rough weather, and probably suggested ultimately the closed decking of the whole of the ship. There is no record of when this feature, which was general in ancient Egyptian vessels, was introduced into Greek galleys. It was certainly in use in the Roman warships about the commencement of the Christian era, for there is in the Vatican a relief of about the date 50 a.d. from the Temple of Fortune at Præneste, which represents part of a bireme, in which the rowers are all below a closed deck, on which the soldiers are standing.

In addition to the fighting-deck proper there were the two side platforms, or gangways, already alluded to, which were carried right round the outside of the vessel on about the same level as the benches of the upper tier of rowers. These platforms projected about 18 to 24 in. beyond the sides of the hull, and were supported on brackets. Like the flying deck, these passages were intended for the accommodation of the soldiers and sailors, who could, by means of them, move freely round the vessel without interfering with the rowers. They were frequently fenced in with stout planking on the outside, so as to protect the soldiers. They do not appear to have been used on galleys of the earliest period.

We have no direct evidence as to the dimensions of ships of four and five banks. Polybios tells us that the crew of a Roman quinquereme in the first Carthaginian War, at a battle fought in 256 b.c., numbered 300, in addition to 120 soldiers. Now, the number 300 can be obtained by adding two banks of respectively 64 and 62 rowers to the 172 of the trireme. We45 may, perhaps, infer that the quinquereme of that time was a little longer than the trireme, and had about 3 ft. more freeboard, this being the additional height required to accommodate two extra banks of oars. Three hundred years later than the above-mentioned date Pliny tells us that this type of galley carried 400 rowers.

We know no detailed particulars of vessels having a greater number of banks than five till we get to the alleged forty-banker of Ptolemy Philopater. Of this ship Callixenos gives the following particulars:—Her dimensions were: length, 420 ft.; breadth, 57 ft.; draught, under 6 ft.; height of stern ornament above water-line, 79 ft. 6 in.; height of stem ornament, 72 ft.; length of the longest oars, 57 ft. The oars were stated to have been weighted with lead inboard, so as to balance the great overhanging length. The number of the rowers was

4,000, and of the remainder of the crew 3,500, making a total of 7,500 men, for whom, we are asked to believe, accommodation was found on a vessel of the dimensions given. This last statement is quite sufficient to utterly discredit the whole story, as it implies that each man had a cubic space of only about 130 ft. to live in, and that, too, in the climate of Egypt. Moreover, if we look into the question of the oars we shall see that the dimensions given are absolutely impossible—that is to say, if we make the usual assumption that the banks were successive horizontal tiers of oars placed one above the other. There were said to have been forty banks. Now, the smallest distance, vertically, between two successive banks, if the oar-ports were arranged as in Fig. 14, with the object of economizing space in the vertical direction to the greatest possible degree, would be 1 ft. 3 in. If the lowest oar-ports were 3 ft. above the water, and the topmost bank were worked on the gunwale, we should require, to accommodate forty banks, a height of side equal to 46 39 ft. × 1 ft. 3 in. + 3 ft. = 51 ft. 9 in. Now, if the inboard portion of the 57 ft. oar were only one-fourth of the whole length, or 14 ft. 3 in., this would leave 57 ft. - 14 ft. 3 in. = 42 ft. 9 in. for the outboard portion, and as the height of gunwale on which this particular length of oar was worked must have been, as shown above, 51 ft. 9 in. above the water, it is evident that the outboard portion of the oar could not be made to touch the water at all. Also, if we consider the conditions of structural strength of the side of a ship honeycombed with oar-ports, and standing to the enormous height of 51 ft. 9 in. above the water-line, it is evident that, in order to be secure, it would require to be supported by numerous tiers of transverse horizontal beams, similar to deck-beams, running from side to side. The planes of these tiers would intersect the inboard portions of many of the tiers of oars, and consequently prevent these latter from being fitted at all.

If we look at the matter from another point of view we shall meet with equally absurd results. The oars in the upper banks of Athenian triremes are known to have been about 14 ft. in length. Underneath them, were, of course, two other banks. If, now, we assume that the upper bank tholes were 5 ft. 6 in.[10] above the water-line, and that one-quarter of the length of the upper bank oars was inboard, and if we add thirty-seven additional banks parallel to the first bank, so as to make forty in all, simple proportion will show us that the outboard portion of the oars of the uppermost bank must have been just under 99 ft. long and the total length of each, if we assume, as before, that one quarter of it was inboard, would be 132 ft., instead of the 57 ft. given by Callixenos. Any variations in the above assumptions, consistent with possibilities, would only have [47]the effect of bringing the oars out still longer. We are therefore driven to conclude, either that the account given by Callixenos was grossly inaccurate, or else that the Greek word, τεσσαρακοντήρης, which we translate by "forty-banked ship," did not imply that there were forty horizontal *superimposed* tiers of oars.

The exact arrangement of the oars in the larger classes of galleys has always been a puzzle, and has formed the subject of much controversy amongst modern writers on naval architecture. The vessels were distinguished, according to the numbers of the banks of oars, as uniremes, biremes, triremes, quadriremes, etc., up to ships like the great galley of Ptolemy Philopater, which was said to have had forty banks. Now, the difficulty is to know what is meant by a bank of oars. It was formerly assumed that the term referred to the horizontal tiers of oars placed one above the other; but it can easily be proved, by attempting to draw the galleys with the oars and rowers in place, that it would be very difficult to accommodate as many as five horizontal banks and absolutely impossible to find room for more than seven. Not only would the space within the hull of the ship be totally insufficient for the rowers, but the length of the upper tiers of oars would be so great that they would be unmanageable, and that of the lower tiers so small that they would be inefficient. The details given by ancient writers throw very little light upon this difficult subject. Some authors have stated that there was only one man to each oar, and we now know that this was the case with the smaller classes of vessels, say, up to those provided with three, or four, to five banks of oars; but it is extremely improbable that the oars of the larger classes could have been so worked. The oars of modern Venetian galleys were each manned by five rowers. It is impossible in this work to examine closely into all the rival theories as to what constituted48 a bank of oars. It seems improbable, for reasons before stated, that any vessel could have had more than five horizontal tiers. It is certain also that, in order to find room for the rowers to work above each other in these tiers, the oar-ports must have been placed, not vertically above each other, but in oblique rows, as represented in Fig. 14. It is considered by Mr. W. S. Lindsay, in his "History of Merchant Shipping and Ancient Commerce," that each of the oblique rows of oars, thus arranged, may have formed the tier referred to in the designation of the class of the vessel, for vessels larger than quinqueremes. If this were so, there would then be no difficulty in conceiving the possibility of constructing galleys with even as many as forty tiers of oars like the huge alleged galley of Ptolemy Philopater. Fig. 15 represents the disposition of the oar-ports according to this theory for an octoreme.

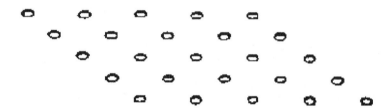

Fig. 14.—Probable arrangement of oar-ports in ancient galleys.

Fig. 15.—Suggested arrangement of oar-ports in an octoreme.

It appears to be certain that the oars were not very advantageously arranged, or proportioned, in the old Greek galleys, or even in the Roman galleys, till the time of the early Cæsars, for we read that the average speed of the Athenian triremes was 200 stadia in the day. If the stadium were equal in length to a furlong, and the working day supposed to be limited to ten hours, this would correspond to a speed of only two and a49 half miles an hour. The lengths of the oars in the Athenian triremes have been already given (p. 42); even those of the upper banks were extremely short—only, in fact, about a foot longer than those used in modern 8-oared racing boats. On account of their shortness and the height above the water at which they were worked, the angle which the oars made with the water was very steep and consequently disadvantageous. In the case of the Athenian triremes, this angle must have been about 23.5°. This statement is confirmed by all the paintings and sculptures which have come down to us. It is proved equally by the painting of an Athenian bireme of 500 b.c. shown in Fig. 9, and by the Roman trireme, founded on the sculptures of Trajan's Column of about 110 a.d., shown in Fig. 16.11 In fact, it is evident that the ancients, before the time of the introduction of the Liburnian galley, did not understand the art of rowing as we do to-day. The celebrated Liburnian galleys, which were first used by the Romans, for war purposes, at the battle of Actium under Augustus Cæsar, 50were said to have had a speed of four times that of the old triremes. The modern galleys used in the Mediterranean in the seventeenth century are said to have occasionally made the passage from Naples to Palermo in seventeen hours. This is equivalent to an average speed of between 11 and 12 miles per hour.

Fig. 16.—Roman galley. About 110 a.d.

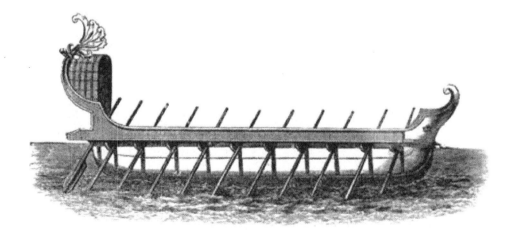

Fig. 17.—Liburnian galley. Conjectural restoration.

The timber used by the ancient races on the shores of the Mediterranean in the construction of their ships appears to have been chiefly fir and oak; but, in addition to these, many other varieties, such as pitch pine, elm, cedar, chestnut, ilex, or evergreen oak, ash, and alder, and even orange wood, appear to have been tried from time to time. They do not seem to have understood the virtue of using seasoned timber, for we read in ancient history of fleets having been completed ready for sea in incredibly short periods after the felling of the trees. Thus, the Romans are said to have built and equipped a fleet of 220 vessels in 45 days for the purpose of resisting the attacks of Hiero, King of Syracuse. In the second Punic War Scipio put to sea with a fleet which was stated to have been completed in forty days from the time the timber was felled. On the other hand, the ancients believed in all sorts of absurd rules as to the proper day of the moon on which to fell trees51 for shipbuilding purposes, and also as to the quarter from which the wind should blow, and so forth. Thus, Hesiod states that timber should only be cut on the seventeenth day of the moon's age, because the sap, which is the great cause of early decay, would then be sunk, the moon being on the wane. Others extend the time from the fifteenth to the twenty-third day of the moon, and appeal with confidence to the experience of all artificers to prove that timber cut at any other period becomes rapidly worm-eaten and rotten. Some, again, asserted that if felled on the day of the new moon the timber would be incorruptible, while others prescribed a different quarter from which the wind should blow for every season of the year. Probably on account of the ease with which it was worked, fir stood in high repute as a material for shipbuilding.

The structure of the hulls of ancient ships was not dissimilar in its main features to that of modern wooden vessels. The very earliest types were probably without external keels. As the practice of naval architecture advanced, keels were introduced, and served the double purpose of a foundation for the framing of the hull and of preventing the vessel from making leeway in a wind. Below the keel proper was a false keel, which was useful when vessels were hauled up on shore, and above the keelson was an upper false keel, into which the masts were stepped. The stem formed an angle of about 70° with the water-line, and its junction with the keel was strengthened by a stout knee-piece. The design of the stem above water was often highly ornate. The stern generally rose in a graceful curve, and was also lavishly ornamented. Fig. 18 gives some illustrations of the highly ornamented extremities of the stern and prow of Roman galleys. These show what considerable pains the ancients bestowed on the decoration of their vessels. There was no rudder-post, the steering having been effected52 by means of special oars, as in the early Egyptian vessels. Into the keel were notched the floor timbers, and the heads of these latter were bound together by the keelson, or inner keel. Beams connected the top timbers of the opposite branches of the ribs and formed the support for the deck. The planking was put on at right angles to the frames, the butting ends of the planks being connected by dovetails. The skin of the ship was strengthened, in the Athenian galleys, by means of stout planks, or waling-pieces, carried horizontally round the ship, each pair meeting together in front of the stem, where they formed the foundations for the beaks, or rams. The hulls were further strengthened by means of girding-cables, also carried horizontally round the hull, in the53 angles formed by the projection of the waling-pieces beyond the skin. These cables passed through an eye-hole at the stem, and were tightened up at the stern by means of levers. It is supposed that they were of use in holding the ship together under the shock of ramming. The hull was made water-tight by caulking the seams of the planking. Originally this was accomplished with a paste formed of ground sea-shells and water. This paste, however, not having much cohesion, was liable to crack and fall out when the vessel strained. A slight improvement was made when the shells were calcined and turned into lime. Pitch and wax were also employed, but were eventually superseded by the use of flax, which was driven in between the seams. Flax was certainly used for caulking in the time of Alexander the Great, and a similar material has continued to be employed for this purpose down to the present day. In addition to caulking the seams, it was also customary to coat over the bottom with pitch, and the Romans, at any rate, used sometimes to sheath their galleys with sheet lead fastened to the planking with copper nails. This was proved by the discovery of one of Trajan's galleys in Lake Riccio after it had been submerged for over thirteen centuries.

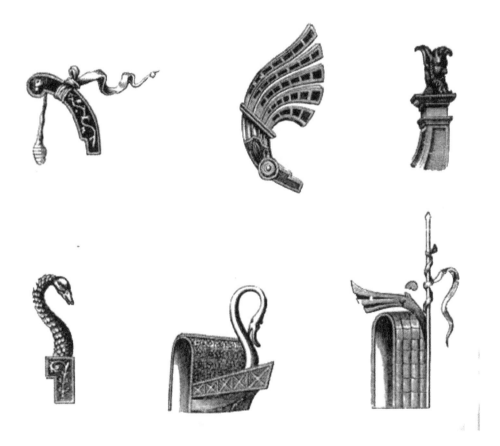

Fig. 18.—Stem and stern ornaments of galleys.

The bows of the ancient war galleys were so constructed as to act as rams. The ram was

made of hard timber projecting   Fig. 19.—Bow of ancient war-galley.
beyond the line of the bow, between it and the forefoot. It was usually made of oak, elm, or
ash, even when all the 54 rest of the hull was constructed of soft timber. In later times it was
sheathed with, or even made entirely of, bronze. It was often highly ornamented, either
with a carved head of a ram or some other animal, as shown in Figs. 8 to 11; sometimes
swords or spear-heads were added, as shown in Figs. 19 and 20. A relic of this ancient

custom is found to this day in the ornamentation   Fig. 20.—Bow of
ancient war-galley. of the prows of the Venetian gondolas. Originally the ram, or rostrum,
was visible above the water-line, but it was afterwards found to be far more effective when
wholly immersed. In addition to the rams there were side projections, or catheads, above
water near the bow. The ram was used for sinking the opposing vessels by penetrating their
hulls, and the catheads for shattering their oars when sheering up suddenly alongside.
Roman galleys were fitted with castles, or turrets, in which were placed fighting men and
various engines of destruction. They were frequently temporary structures, sometimes
consisting of little more than a protected platform, mounted on scaffolding, which could be
easily taken down and stowed away. The use of these structures was continued till far into
the Middle Ages.

CHAPTER III.

ANCIENT SHIPS IN THE SEAS OF NORTHERN EUROPE.

Outside the Mediterranean it is known that some of the northern nations had attained to very considerable skill in the arts of shipbuilding and navigation. Cæsar gives a general description of the ships of the Veneti, who occupied the country now known as Brittany, and who had in their hands the carrying trade between Gaul and Britain.12 As might be expected from the stormy nature of the Atlantic, the Veneti were not able to place any reliance on oars as a means for propulsion. According to Cæsar's account, they trusted solely to sails. Their vessels were built entirely of oak of great thickness. He also mentions that the beams were as much as 12 in. in depth. The bottoms of these vessels were very flat, so as to enable them the better to be laid up on the beach. The hulls had considerable sheer, both at the stem and stern. The sails were of dressed hide, and the cables were iron chains. It is evident from this cursory description that the ships of the Veneti were not based upon Mediterranean models, and it is highly probable that they, rather than the oar-propelled galleys, may be regarded as the prototypes of the early sea-going vessels of Northern Europe.

Although the art of ship construction had attained to great importance amongst the Veneti, their neighbours, the Britons, were still very backward in this respect at the time of the first Roman invasion. Cæsar states that their vessels were of very 56slight construction, the framework being made of light timber, over which was stretched a covering, or skin, of strong hides. Sometimes the framework was of wicker.

The ancient Saxons, who were notorious as pirates on the North Sea, made use of boats similar to those of the ancient Britons. At the time of their invasion of Britain, however, their vessels must have been larger and of more solid construction, though we must dismiss, as an obvious absurdity, the statement that the first invading army of 9,000 men was carried to this country in three ships only. It is much more probable that the expedition was embarked in three fleets.

The Saxon kings of England often maintained very considerable fleets for the purpose of protecting the coast from the Danes.

Alfred the Great is generally regarded as the founder of the English Navy. He designed ships which were of a better type and larger size than those of his enemies, the Danes. They were said to have been twice as long as the vessels which they superseded. The Saxon Chronicle says, "They were full twice as long as the others; some had sixty oars, and some had more; they were swifter and steadier, and also higher than the others; they were shaped neither like the Frisian, nor the Danish, but so as it seemed to him they would be most efficient." In 897 Alfred met and defeated a Danish squadron, in all probability with his new ships.

Edgar (959 to 975) is stated to have kept at sea no less than 3,600 vessels of various sizes, divided into three fleets, and the old historian William of Malmesbury tells us that this king took an active personal interest in his navy, and that in summer time he would, in turn, embark and cruise with each of the squadrons.

Fig. 21.—Anglo-Saxon ship. About 900 a.d.

Fig. 21 is an illustration of an Anglo-Saxon ship taken from an old Saxon calendar, which is, or was, in the Cottonian57 Library, and which is supposed to have been written about half a century before the Norman Conquest. It is reproduced in Strutt's "Compleat View of the Manners, Customs, Arms, Habits, etc., of the Inhabitants of England, from the arrival of the Saxons till the reign of Henry VIII.," published in 1775. The proportions of the boat as represented are obviously impossible. The sketch is, however, interesting, as showing the general form and mode of planking of the vessel, and the nature of the decorations of the bow and stern. We see that the vessel was a warship, as the keel prolonged formed a formidable ram. We also may notice that the sail was relied on as a principal means of propulsion, for there are apparently no notches or rowlocks for oars. The steering was effected by two large oars, in a similar manner to that adopted by the ancient Egyptians and other Mediterranean peoples. The extraordinary character of the deck-house will be observed.58 It is, of course, purely symbolical, and may, at most, be interpreted as meaning that the vessel carried some sort of structure on deck.

In the seventh and eighth centuries of the Christian era the scene of maritime activity was transferred from the Mediterranean to the North of Europe. The Norsemen, who overran

44

the whole of the European seaboard at one time or another, were the most famous navigators of the period immediately preceding the Middle Ages. Any record connected with their system of ship-construction is necessarily of great interest. The fleets of the Norsemen penetrated into the Mediterranean as far as the imperial city of the Eastern emperors. In the north they discovered and colonized Iceland, and even Greenland; and there are good grounds for believing that an expedition, equipped in Iceland, founded a colony in what are now the New England States five centuries before Columbus discovered the West Indies. Unfortunately, the written descriptions extant of the Norse ships are extremely meagre, and if it had not been for the curious custom of the Norsemen of burying their great chiefs in one of their ships and heaping earth over the entire mass, we should now know nothing for certain of the character of their vessels. Many of these ship-tombs have been discovered in modern times, but it happened in the majority of instances that the character of the earth used was unsuited to their preservation, and most of the woodwork was found to be decayed when the mounds were explored. Fortunately, however, in two instances the vessels were buried in blue clay, which is an excellent preserver of timber, and, thanks to the discovery of these, we have now a tolerably complete knowledge of the smaller classes of vessels used by the Vikings. One of them was discovered, in 1867, at Haugen, but by far the most important was found in 1880, at Gogstad, near Sandefjord, at the entrance of the Fjord of Christiania.59 Though this vessel is comparatively small, she is, probably, a correct representative of the larger type of ships made use of by the renowned adventurers of the North in their distant expeditions.

In view of the great interest attaching to this find, a detailed description of the vessel is given. The illustrations (Figs. 22 to 26), showing an end elevation, longitudinal and cross-sections, and the half-plan with her lines, are taken from the "Transactions of the Institution of Naval Architects."13 The boat was clinker-built and wholly of oak. Her principal dimensions are: length, 77 ft. 11 in.; extreme breadth, 16 ft. 7 in.; and depth, from top of keel to gunwale, 5 ft. 9 in. The keel is 14 in. deep, the part below the rabbet of the garboard or lowest strakes of the planking, being 11 in. deep, and 4½ in. thick at the bottom. The width across the rabbet is 3 in., while the portion above the rabbet and inboard is 7 in. wide. The keel and stem and stern-posts run into each other with very gentle curves. The keel itself is 57 ft. long, and to it are connected, by vertical scarves and a double row of iron rivets, the forefoot and heel-pieces, which latter are fastened in a similar manner to the stem and stern-post. These posts are 15 in. deep at the scarf, gradually tapering upwards. The framing of the bottom is formed of grown floors resting on the top of the keel, and extending in one piece, from shelf to shelf, as shown in the transverse section (Fig. 23). There are nineteen of these floors in all, spaced in the body of the boat, on the average 3 ft. 3 in. apart. They are 4 in. in diameter at the garboard strake, and taper in both dimensions, so that they are less than 3 in. at the shelf. They are not fastened to the keel. The planking is put on clinker fashion. There are sixteen strakes a side, the breadth of each, amidships, being on the average 9½ in., including the land of 1 in., and 6160the length of planks varies from 8 ft. to 24 ft. The thickness is generally 1 in. The tenth plank from the keel is, however, 1-3/4 in. thick, and forms a kind of shelf for the beam-ends. The third plank from the top is 1¼ in. thick, and is pierced with 4-in. holes for the oars, of which there are sixteen on each side. The two upper strakes are only 3/4 in. thick, and inside the top one is placed the gunwale, which is 3 × 4½. The planks are fastened together by iron rivets spaced from 6 in. to 8 in. apart. The heads of the rivets are 1 in. in diameter, and the riveting plates 1/2 in. square. The planks are worked down from thicker slabs, and a ledge 1 in. in height is left on the inboard surface of the middle of each plank. The planks bear against each floor at two points, viz. the upper edge and the projecting ledge. Fig. 24 shows a section of a floor and of the plank, with its projecting ledge. The fastenings of the planking to the floors are very peculiar. Two holes are bored transversely in the ledge, one on either side of each floor. There is a corresponding hole running fore and aft through the floor, and through these holes are passed ties made of the tough roots of trees barely 1/4 in. in diameter, crossed on the ledge and passing once through each hole. The only iron fastening between the planking and the floors is at the extreme ends of the latter, where a single nail is driven through each, and riveted at the ends of the floors. The beams rest on the shelf strake and on the tops of the floor-ends. They are 7 in. deep and 4 in. wide. They are connected with the planking by knees (see the section, Fig. 23), fastened to their upper faces and to the side of the ship as far up as the oar-strake, or "mainwale," by means of oak trenails. The knees are not so wide as the beams, and consequently a ledge, or landing, is left on each side of the latter which supports the flooring, or bottom boards. The top strakes are connected to the body of the vessel by short timbers, shown in the section, Fig. 23.62 These are placed in the spaces between the knees. The beams are supported in the middle by short pillars resting on the throats of the floors.

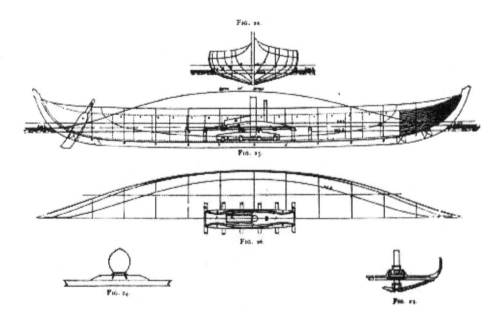

Fig. 22.—Viking ship.

The vessel was propelled by sails as well as oars. It was fitted with a single mast; the arrangements for stepping and raising and lowering the latter were peculiar. A beam of oak, 11 ft. long, 19 in. wide, and 14 in. deep, formed the step. A side elevation of this is shown at *s*, in the longitudinal section, Fig. 25, and a cross-section in Fig. 23. The step, as may be seen, is countersunk over the throats of the floors; it is tapered towards the ends, and a piece (*c*) nearly 12 in. thick, immediately forward of the mast, rises vertically out of it. This piece is fastened to a huge log of oak, 16 ft. long, 38 in. broad, and 14 in. deep in the

46

middle, marked *I* (Figs. 25 and 26), which rests on a sole-piece about 4 in. thick. The sole-piece is countersunk over the beams. The large log is called by Mr. Colin Archer the "fish," partly because its ends are fashioned to represent the tails of two whales, and partly because the mast partners of modern ships, which take the place of this heavy piece, are to this day called *Fisken* in Norway. The fish contains a slot (*h*) nearly 6 ft. long, and the same width as the mast, 12½ in. The mast goes through the forward end of the slot, and when it is in use the slot is filled up with a heavy slab. When the mast is lowered for going into action, or when going against a head-wind, the slab is removed, and the fore-stay slacked off, thus permitting the mast to fall aft. The sail used was a solitary square one. The rudder resembles a short oar. It is hung by a rope passing through a perforated conical chock on the starboard side of the ship. There is an iron eyebolt near the bottom edge, through which a rope probably passed for the purpose of raising the rudder when not in use. The rudder was worked by means of a tiller fitted into the socket at the upper end.

63Unfortunately, the two extreme ends of the ship have decayed away, so that it is not possible to determine with accuracy what was the appearance of the bow and stern. It is, however, probable, from the direction taken by the planking towards the ends, that the vessel possessed very considerable sheer. As may be seen from the plan, the character of the lines was extremely fine, and it is probable that the boat was capable of high speed. The remains of the ropes which have been discovered prove that they were made from the bark of trees.

This vessel may be considered as a connecting link between the ancient and mediæval types of ships. Her proportions and scantlings prove that her builders had a large experience of shipbuilding, that they fully understood how to work their material and to adapt it properly to the duty it had to fulfil, and also that they understood the art, which was subsequently lost, to be revived only in modern times, of shaping the underwater portion of the hull so as to reduce the resistance to the passage of the vessel through the water. The only part of the structural design to which any serious exception can be taken is the very slight character of the connection between the top sides and the body of the boat, and even this defect was probably not very serious when we take into account the lightness of the loading, and the fact that it probably consisted chiefly of live cargo, so that there was little dead weight to cause serious straining.

Vessels of the type of the Viking ships were built in Denmark at a very early date. In 1865 three boats were discovered buried in a peat bog in Jutland. Danish antiquaries consider that they were built about the fifth century of our era. The largest is 70 ft. in length and of such an excellent type that boats of somewhat similar form and construction are in universal use to this day all round the coasts of Norway.64 Such an instance of persistency in type is without parallel in the history of shipbuilding, and is a wonderful proof of the skill of the Norsemen in designing and building vessels. The boat in question is clinker-built, the planks having the same peculiarities as those of the Viking ship just described. It is of the same shape at both ends, and has great sheer at both stem and stern. The rowlocks, of which there are thirty, prove that the vessel was intended to be rowed in either direction. This also is a peculiarity of the modern Norwegian rowboat. The steering was effected by means of a large oar, or paddle. There is no trace of a mast, nor of any fitting to receive one; nor was the vessel decked. The internal framing was admirably contrived. In fact, it would be difficult, even at the present time, to find a vessel in which lightness and strength were better combined than in this fifteen-hundred-year-old specimen of the shipbuilder's art.

CHAPTER IV.

MEDIÆVAL SHIPS.

In the times of the Norman kings of England both the war and the mercantile navies of the country were highly developed. William the Conqueror invaded this island without the assistance of a war navy. He trusted to good luck to transport his army across the Channel in an unprotected fleet of small vessels which were built for this purpose, and which were burnt by his order when the landing had been effected. We possess illustrations of these transport vessels from a contemporary source—the Bayeux tapestry, which was, according to tradition, the work of Queen Matilda, the Conqueror's consort. Fig. 27 represents one of these vessels. It is obviously of Scandinavian type, resembling in some of its features the Viking ship shown in Figs. 22 to 26. Apparently, oars were not used in this particular boat; the propulsion was effected by means of a single square sail. The mast unshipped, as we know from other illustrations on the same piece of tapestry. The steering was effected by a rudder, or steering-board, on the starboard-side. In all the illustrations of ships in this tapestry the main sheet was held by the steersman, a fact which shows that the Normans were cautious navigators. Another ship is represented with ten horses on board.

We possess confirmatory evidence that the ship shown in Fig. 27 represents a type that was prevalent on our coasts in the eleventh and two following centuries, for very similar boats are shown in the transcript of Matthew Paris's "History66 of the Two Kings of Offa" (now in the Cottonian Library), the illustrations in which are supposed to have been drawn by Matthew Paris himself. The history is that of two Saxon princes who lived in the latter half of the eighth century, and was written in the first half of the thirteenth. We may fairly suppose that the illustrations represented the types of vessels with which the historian was familiar. They were all of the type depicted in the Bayeux tapestry. They are of the same shape at both ends, just like the Viking ship, and it may be added, like the boats to this day in common use along the coasts of Norway.

Fig. 27.—One of William the Conqueror's ships. 1066 a.d.

It must not be supposed that the art of building ships of larger size, which was, as we have seen, well understood by the Romans, about the commencement of our era, was forgotten. On the contrary, though, no doubt, the majority of ships of the eleventh and twelfth centuries were of small dimensions, yet we occasionally meet with notices of vessels of comparatively67 large size. Such an one, for instance, was *La Blanche Nef,* built in the reign of Henry I., and lost on the coast of Normandy in the year 1120 a.d. This ship was built for Prince William, the son of the King, and he was lost in her, together with 300 passengers and crew. This number proves that the vessel was of considerable size. *La Blanche Nef* was a fifty-oared galley. Long before her time, at the end of the tenth century, when Ethelred the Unready was King of England, the Viking Olaf Tryggvesson built, according to the Norwegian chroniclers, a vessel 117 ft. in length.

It may here be mentioned that galleys continued to be used, along with sailing ships, in the various European navies till the seventeenth century.

Another instance of the loss of a large twelfth-century ship occurred in the reign of Henry II., half a century later than the wreck of *La Blanche Nef,* when a vessel engaged in transport work foundered with 400 persons.

In the reign of Richard Cœur de Lion a great impetus was given to shipbuilding and to maritime adventure in this country by the expedition which the king undertook to the Holy Land. A fleet of about 110 vessels, according to Peter Langtoft, sailed from Dartmouth in April, 1190 a.d. It was reinforced considerably in the Mediterranean; for, according to Matthew Paris, Richard was accompanied on his voyage to Palestine by 13 buccas, 100 "ships of burthen," and 50 triremes, and according to Vinesauf, the fleet consisted of about

230 vessels. The buccas, or busses, or dromons, were ships of the largest size, with triple sails. There were two sorts of galleys; some were propelled by oars alone, and others by oars and sails: the latter were the larger, and, according to Matthew Paris, sometimes carried 60 men in armour, besides 104 rowers and the sailors. He also states that some of them had triple banks of oars like the ancient galleys; but, according to Vinesauf,68 the majority had not more than two banks of oars, and carried the traditional flying deck above the rowers for the use of the soldiers; they were low in the water compared to the sailing-ships, and they carried beaks, or rams, which, as narrated subsequently, they used to some purpose.

The larger type of sailing-ships carried a captain and fifteen sailors, forty knights with their horses, an equal number of men-at-arms, fourteen servants, and complete stores for twelve months. There were, moreover, three much larger vessels in the fleet which carried double the complement mentioned above.

As an instance of the very large size to which vessels occasionally attained in those days in the Levant, we may refer to a Saracen vessel which was attacked by Richard's fleet near Beirut in Syria, in 1191. It was described by many of the old chroniclers. This ship had three masts, and is alleged to have had 1,500 men on board at the time of the fight. The attack was carried out with great difficulty, on account of the towering height of the sides of the Saracen vessel, and it was not till ramming tactics were tried by the galleys charging in line abreast, that her hull was stove in, in several places, and she went down with nearly all hands, only thirty-five, or, according to other accounts forty-six, having been saved.

These large ships appear to have been used by other Mediterranean Powers towards the end of the twelfth century. For instance, a great Venetian ship visited Constantinople in 1172 a.d., of which it was stated that "no vessel of so great a bulk had ever been within that port." This vessel is mentioned by Cinnamis, Marino, and Filiasi, and others, but her dimensions are not given. It is, however, known that she had three masts. Cinnamis, who was at Constantinople at this very time, states that she received from 1,500 to 2,000 Venetian refugees on board, and conveyed them to the Adriatic. The Venetians are69 said to have employed another very large ship at the siege of Ancona in 1157 a.d. On account of its size it was named *Il Mondo*.

The Republic of Venice was, during the time of which we are writing, and for a long subsequent period, the foremost maritime power of the world. It is highly probable that many of the improvements which found their way into mediæval ships owed their origin to its great naval arsenal, which was famed for its resources and for the technical skill of its employés. At one time this arsenal employed 16,000 workmen, and during the great struggle of the Republic with the Turks at the end of the sixteenth century it turned out a completed and fully equipped galley every day for a hundred days in succession. During the Crusades, Venice and the rival Republic of Genoa secured between them the great bulk of the business involved in transporting troops and stores to the East, and they frequently hired out their war and merchant ships to other Powers.

Shortly after the Crusade of Richard Cœur de Lion the trade and shipping of England appear to have undergone great expansion. In the reign of Henry III. (1216 to 1272) the historian, Matthew of Westminster, writes of them in a strain which might almost apply to our own day:—

"Oh England, whose antient glory is renowned among all nations, like the pride of the Chaldeans; the ships of Tarsis could not compare with thy ships; they bring from all the quarters of the world aromatic spices and all the most precious things of the universe: the sea is thy wall, and thy ports are as the gates of a strong and well-furnished castle."

In another place the same historian writes of the English trade as follows:—

"The Pisans, Genoese, and Venetians supply England with the Eastern gems, as saphires, emeralds, and carbuncles; from Asia was brought the rich silks and purples; from Africa the cinnamon and balm; from Spain the kingdom was enriched with gold; with silver 70from Germany; from Flanders came the rich materials for the garments of the people; while plentiful streams of wine flowed from their own province of Gascoigny; joined with everything that was rich and pretious from every land, wide stretching from the Hyades to the Arcturian Star."

No doubt this expansion was due, in part, to the very large participation which the English fleet took in the Crusade. Great numbers of English mariners were thus enabled to penetrate into seas that were new to them, and had opportunities of studying the commercial needs of the countries which bordered on those seas. Another cause which powerfully contributed to the development of navigation, and consequently of shipbuilding, was the introduction of the mariner's compass into Western Europe during the first half of the thirteenth century.

The English war navy, also at the commencement of the reign of Henry II., appears to have been in a very efficient condition. Matthew Paris gives a description of a great naval fight off the South Foreland, in the year 1217, between a Cinque Ports Fleet under the famous Hubert de Burgh, who was at the time Governor of Dover Castle, and a large French fleet under a monk of the name of Eustace, who was one of the most skilful naval commanders of his day. The English fleet consisted of forty vessels, of which only sixteen were large and manned with trained sailors. The French fleet, which was endeavouring to carry a strong invading army to England, was made up of eighty large vessels, besides numerous galleys and smaller craft. The account of the battle is most interesting, because it throws a flood of light upon the naval tactics and the weapons of offence of the day. The English commander manœuvred for the wind, and having got it, he bore down on the French fleet, and attacked their rear ships with flights of arrows carrying phials of unslaked lime, which being scattered and carried by the wind, blinded the Frenchmen;71 boarding was then attempted with perfect success, the rigging and halyards of the French ships were cut away, causing the sails to fall upon their crews. A hand-to-hand combat then took place, which resulted in fearful slaughter of the would-be invaders: several of the French ships were rammed and sunk by the English galleys, and in the end the whole of the hostile fleet, with the exception of fifteen vessels, was taken or sunk. This was one of the most momentous naval battles in English history, and is memorable as having furnished the first recorded instance of a battle having been preceded by manœvres to obtain the weather-gauge.

Fig. 28. Sandwich seal. 1238.

Fig. 29.—Dover seal. 1284.

We have, unfortunately, very few illustrations of the thirteenth-century ships, and those which we do possess are taken from the corporate seals of some of the Cinque Ports and other southern seaport towns. Fig. 28 is a representation of the seal of Sandwich, and dates from the year 1238. The[72] circular form of a seal is not very favourable for the representation of a masted ship, but we can at least make out that the vessel in question is of the Scandinavian type used by William I. and his successors. It also appears to have been an open boat, and contains the germs of the castellated structures fore and aft, which, as we shall see afterwards, attained to the most exaggerated dimensions. In the case of the Sandwich ship these castles were not incorporated with the structure of the vessel; they were merely elevated positions for the use of the archers and men-at-arms, and were mounted on columns, and were probably removable. We can also learn from the engraving that the practice of furling sails aloft was practised at that time. Fig. 29 is the seal of Dover, and dates from the reign of Edward I. (1284 a.d.). It does not show much progress over the Sandwich boat of nearly fifty[73] years earlier, but we may notice that the castles are more developed and of a more permanent character. This vessel also possesses a bowsprit.

It was about the middle of this century that cabins appear to have been introduced into English ships. The first mention of them occurs in 1242, when orders were given that "decent chambers" were to be constructed in a ship in which the king and queen were to voyage to Gascony.

There are records in existence of the dimensions of some vessels which were built for Louis IX. of France in the year 1268 a.d. at Venice and Genoa. They are published in Jal's "Archéologie Navale." The Venetian ship which was named the *Roccafortis* appears to have been the largest. Her dimensions are given as follows: length of keel, 70 ft.; length over all, 110 ft.; width at prow and poop, 40 ft. This latter dimension is hardly credible. The *Roccafortis* had two covered decks, and a castle or "bellatorium" at each end, and also several cabins. The crew numbered 110.

The Genoese ships were smaller. Two of them were of identical dimensions, viz. length of keel, 49½ ft.; length over all, 75 ft.; beam, 10 ft. The figure given for the beam appears to be too small in this case, if the dimensions of the mast, 70½ ft., are correct, for such a long mast could hardly have been carried in so narrow a boat. These vessels had two decks, and are said to have had stabling for fifty horses each; but this latter statement cannot be true if the dimensions are accurately given.

We have very little information about the ships of the end of the thirteenth and commencement of the fourteenth centuries. There is a list in existence of Cinque Ports ships which were fitted out in 1299 to take part in the war against Scotland. They were thirty in number. More than half of them had complements of two constables and thirty-nine mariners,[74] and the smallest had one constable and nineteen mariners. There is also a statement of the tonnage and complements of ships intended for an expedition to Guienne in the year 1324, which throws some light on the size of the vessels employed in the Scottish expedition. From it we learn that a ship of 240 tons had 60 mariners and officers; one of 200 tons, 50; vessels between 160 and 180 tons, 40; of 140 tons, 35; of 120 tons, 28; of 100 tons, 26; of 80 tons, 24; and of 60 tons, 21. From the above we may infer that the largest vessels in the Cinque Ports' squadron of 1,299 were from 160 to 180 tons. The measure of a ton in those early days was probably the cubic space occupied by a tun of wine of 252 gallons in the hold of a ship.

We possess one representation of an English ship of the date of this expedition to Guienne. It was engraved on the seal of the Port of Poole in the year 1325 (Fig. 30). It is remarkable as the earliest known instance of an English ship fitted with a rudder at the stern instead of the side-rudder, or paddle, which had been in use from the very earliest times. We also notice in this ship a further development of the stern and forecastles, which, however, were not as yet fully incorporated with the structure of the hull.

The reign of Edward III., which commenced in 1327, was, in consequence of the wars with Scotland and France, one of great naval activity. After some years of desultory naval warfare in the Channel, a famous sea fight took place at Sluys, in Dutch Flanders, about ten miles north-east of Blankenberghe, in the year 1340. The English fleet consisted of about 200 ships under the personal command of Edward III. The allied French and Genoese fleet numbered, according to the English king, 190, and was composed of ships, galleys, and barges, while some of the chroniclers have put its numbers at as many as 400 sail, but this would probably include many small craft.75 The battle resulted in the capture, or destruction, of nearly the whole French fleet. The English are said to have lost 4,000 men killed, and the French 25,000. In one vessel, named the *Jeanne de Dieppe*, captured by the Earl of Huntingdon, no fewer than 400 dead bodies were found. The latter figure shows that some very large vessels were used at this battle.

Fig. 30.—Poole seal. 1325.

55

Edward III. caused a gold noble to be struck in 1344 bearing the representation of a ship almost precisely similar to the vessel on the seal of Poole, of about twenty years earlier (Fig. 30). It is fitted with a rudder at the stern, and we may therefore conclude that at this period the side-rudder, or clavus, had disappeared from all important vessels. The fore and stern castles were, in most cases, temporary additions to merchant ships, to adapt them for purposes of warfare. In fact, nearly all the sailing-ships used in naval warfare down to, and even after the fourteenth century, appear to have been employed as merchant vessels in time of peace; and this76 remark applies even to the king's ships. It was, no doubt, the introduction of artillery that first caused the sailing warship to be differentiated from the merchantman. Although gunpowder for military purposes is said to have been used on land as early as 1326, and although iron and brass cannon are mentioned amongst the stores of three of the king's ships in 1338, nevertheless, the battle of Sluys and the subsequent naval engagements in the reign of Edward III. appear to have been fought without artillery. It was not till the last quarter of the fourteenth century that guns became at all common on board ship.

In the year 1345 Edward III. invaded France, and was accompanied by a fleet of from 1,000 to 1,100 ships, besides small craft. Two hundred of these vessels were employed after the king's landing in ravaging the northern coasts of France and destroying the hostile shipping.

In the year 1347 Edward organised another great naval expedition against France, this time in order to give him the command of the sea during his siege of Calais. The fleet was drawn from all the ports of the kingdom, and small contingents came from Ireland, Flanders, Spain, and the king's own possession of Bayonne. There are two lists in existence of the numbers of ships and men contributed by each port to this expedition. They agree very closely. According to one of them, the united fleet consisted of 745 ships, and 15,895 mariners, or an average of about twenty mariners to each ship. This figure, of course, does not include the fighting men. About fifty of these vessels were fighting ships fitted with castles, and the remainder were barges, ballingers (which appear to have been a kind of large barge), and transports. The largest contingents, by far, came from Yarmouth, which contributed 43 ships and 1,950 men; Fowey sent 47 ships and 770 men; and Dartmouth supplied 32 ships and 756 men;77 while London, independently of the king's own vessels, sent only 25 ships manned with 662 men.

In 1350 Edward III. and the Black Prince fought a famous naval battle off Winchelsea against a fleet of forty Spanish ships. The battle is generally known by the name of L'Espagnols-sur-Mer. Edward was victorious, though he lost his own ship, through its springing a leak when colliding with one of the Spanish vessels. The tactics of the English consisted chiefly of boarding, while the Spaniards, whose vessels were much the higher, attacked with cross-bows and heavy stones; the latter they hurled from their fighting-tops into their adversaries' ships.

From the foregoing, we can infer that the naval resources of England in the first half of the reign of Edward III. were very great. During the latter half of his reign he neglected his navy, and the French and Spaniards, in spite of all their previous losses, rapidly gained the upper hand at sea, and ravaged the English coasts. In 1372 the Spanish fleet assisting the

French inflicted a severe defeat upon an inferior English squadron which had been sent to the relief of La Rochelle. This battle is memorable because it was, probably, the first sea-fight in which artillery was employed, the Spanish ships having been partly armed with the new weapon. The Venetians are usually credited with having been the first people to employ naval guns; but we do not find them using artillery against the Genoese till the year 1377.

The introduction of cannon as the armament of ships of war was the cause of several modifications in the construction of their hulls. Most of the early vessels fitted with cannon were of the galley type, the guns being mounted on the upper deck, and fired over the bulwarks, *en barbette.* Afterwards portholes were cut through the bulwarks. Fig. 31 represents a Venetian galley of the fourteenth century, as given by Charnock, with a single gun mounted in the bow.

Fig. 31.—Venetian galley. Fourteenth century.

78The new form of armament of ships involved a considerable raising of the height of side, and in order to counteract the effect of the high topside, carrying the weight of guns aloft, the beam of the vessel relatively to its length had to be much increased. The Venetians were, however, afraid to make the transverse section wide throughout, lest the weight of the guns near the sides of the vessel should cause the connection of the sides with the beams to strain; hence they gave the sides considerable "tumble home," or fall inboard, as represented by Fig. 32, which shows the cross-section of a Venetian galleon. It will be noticed that the width of the upper deck is only about half that of the greatest beam. This practice was afterwards carried to an absurd extent by the Venetians and their imitators, even in cases where guns were not carried aloft, as may be seen from the sketch of a galleon given in Fig. 33. Hence it is evident that the introduction of ordnance on board ship accounted for a complete revolution in the proportions of hulls hitherto in vogue. The rig of ships also underwent a79 considerable development about this period. The old single mast of the galley was supplemented by two and in some cases by three others. The sails were still square sails carried on spars, and the practice of reefing the sails to the spars aloft, instead of lowering spars and sails together on deck, had now become common.

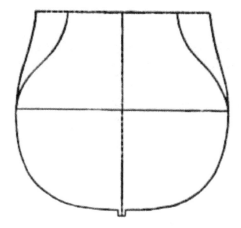

Fig. 32.—Cross-section of a Venetian galleon.

Two years after the action off La Rochelle we find the French commencing the construction of a Royal Navy at Rouen. This step was taken in consequence of the strong opinion held by Jean de Vienne, who was appointed Admiral of France in 1373, that vessels built specially for the purposes of war would have a great advantage over the hired merchantmen which had to be adapted for fighting each time they were impressed.

It is highly probable that the latter half of the fourteenth century witnessed many improvements in ships built in the Mediterranean. This was no doubt due, in part, to the intense commercial rivalry that existed at that time between Venice and the other Italian Republics. Fig. 34 is taken from a MS. Virgil in the Riccardi Library, reproduced in M. Jal's14 work. It represents an Italian two-masted sailing-ship of this period. This is one of the earliest illustrations of a ship 80with a permanent forecastle forming part of the structure of the vessel. The stern castle also appears to have a permanent, though not a structural character. Ships of somewhat similar type were used in England in the reign of Richard II. at the end of the fourteenth century. Fig. 35 represents one of them, the original being in an illustrated manuscript in the Harleian Library. It was written by a Frenchman of the name of Francis de la Marque in Richard's reign. There are illustrations in manuscripts still in existence written about this period, which confirm the fact that this type of ship was then prevalent.

Fig. 33.—Venetian galleon. 1564.

The reign of Henry V. (1413 to 1422) was one of great naval development. The king himself took a most ardent interest in the Royal Navy, and frequently inspected the ships during their construction. Under his auspices some very large vessels were built for the fleet. Lists of this king's ships are81 still in existence. They are classified under the names Great Ships, Cogs, Carracks, Ships, Barges, and Ballingers. The largest of the great ships was the *Jesus*, of 1,000 tons; the *Holigost*, of 760; the *Trinity Royal*, of 540; and the *Christopher Spayne*, of 600; the last-mentioned was a prize captured by the Earl of Huntingdon. The majority of the ships were, however, from 420 to 120 tons. The carracks were apparently not English-built ships, as all those in the king's navy were prizes captured in 1416 and 1417. The three largest were of 600, 550, and 500 tons respectively. The barges are given as of 100 tons, and the ballingers ranged from 120 to 80 tons. The total strength of the Royal Navy about the year 1420, as given in the list compiled by82 W. M. Oppenheim from the accounts of the keepers of the king's ships, is 38; of these 17 were ships, 7 carracks, 2 barges, and 12 ballingers. It is worthy of notice that there were no galleys included in the list.

Fig. 34.—Italian sailing ship. 15th century.   Fig. 35.—English ship. Time of Richard II.

Henry invaded France in 1415 with a fleet of 1,400 vessels, which had been raised by impressing every British ship of 20 tons and upwards. The home supply not being sufficient for his purpose, Henry sent commissioners to Holland and Zealand to hire additional vessels. In all 1,500 ships were collected and 1,400 utilised. These figures give us a fair idea of the resources of this country in shipping at that time.

This was the invasion which resulted in the victory of Agincourt and the capture of Harfleur. In the year following (1416) France was again invaded and the fleet was stated by some to have numbered 300, and by others 400 ships. A naval battle was fought off Harfleur. It resulted in a complete victory for Henry. The old tactics and the old weapons seem to have been used. Although, as we have seen, guns had been used in sea-fights nearly forty years previously, there is no mention of their having been employed on either side at this battle.

In 1417 the king again collected 1,500 vessels at Southampton for a fresh invasion of France. Having first obtained the command of the sea by a naval victory over the French and Genoese, a landing was duly effected near Harfleur. Several vessels, including four large carracks, were captured in the sea-fight, and were added to the king's navy.

During the reign of Henry V. the Mercantile Marine of England made no progress. Commerce was checked in consequence of the state of war which prevailed, and the improvements in shipbuilding seem to have been confined to the Royal Navy. It seems probable, however, that the experience gained in the construction and navigation of the 83 very large ships which the king added to the navy had its effect, ultimately, in improving the type of merchant-vessels.

Fig. 36.—English ship. Time of Henry VI.

During the forty years of the reign of Henry VI. England was so greatly exhausted and impoverished by war with France and by internal dissensions at home, that commerce and shipbuilding made little progress. We possess a sketch of a ship of the early part of the reign of Henry VI. It is contained in a manuscript in the Harleian Library of the date, probably, of 1430 to 1435. It is reproduced in Fig. 36, and differs from the ship of the reign of Richard II. shown in Fig. 35, chiefly in having the poop and forecastle more strongly developed.

While England was steadily declining in power from the time of the death of Henry V., a new maritime nation was arising in South-Western Europe, whose discoveries were destined to have a most marked effect on the seaborne commerce, and consequently on the shipbuilding of the world. In the year 1417 the Portuguese, under the guidance of Prince Henry the Navigator, commenced their exploration of the west coast of 84 Africa, and they continued it with persistency during the century. In 1418 they discovered, or rather re-discovered, the island of Madeira, for it is extremely probable that it was first visited by an Englishman of the name of Machin.

The Portuguese prince firmly believed that a route could be opened round Africa to the Indies. To reach these regions by sea seems to have been the goal of the great explorers of the fifteenth century, and the Portuguese were stimulated in their endeavours by a grant from Pope Martin V. of all territories which might thenceforward be discovered between Cape Bojador and the East Indies. In 1446 an expedition consisting of six caravels was fitted out, and made a voyage to Guinea; it resulted in the discovery of the Cape Verde Islands. The caravel was a type of ship much used by the countries of Southern Europe in the fifteenth and sixteenth centuries. A description of a Spanish vessel of this type is given on pages 87 to 89. In 1449 the Azores were discovered. In 1481 a lucrative trade was opened up between Portugal and the natives of Guinea. Six years afterwards the Cape of Good Hope was reached by Bartholomew Diaz, and in 1497 it was doubled by Vasco da Gama.

During a great part of the period in which the Portuguese were thus occupied in extending their commerce and in paving the way for great discoveries, the condition of England, owing to the French war and to the subsequent Wars of the Roses, was passing from bad to worse. Nevertheless, the spirit of commercial enterprise was not wholly extinguished. A few merchants seem to have made fortunes in the shipping trade, and among them may be mentioned the famous William Canynge of Bristol, who was probably the greatest private shipowner in England at the end of the reign of Henry VI. and during the time of Edward IV. (1461 to 1483). Canynge traded to Iceland, Finland, and the Mediterranean. He is85 said to have possessed ships as large as 900 tons, and it is recorded on his monument, in the church of St. Mary Redcliffe, in Bristol, that he at one time lent ships, to the extent of 2,670 tons, to Edward IV. It is also related of him that he owned ten ships and employed 800 sailors and 100 artisans.

It was not till the year 1475, upon the conclusion of peace between Edward and the French king, Louis, that affairs quieted down in England, and then trade and commerce made most marvellous progress. The king himself was one of the leading merchants of the country, and concluded treaties of commerce with Denmark, Brittany, Castile, Burgundy, France, Zealand, and the Hanseatic League. In the reign of Edward's successor, Richard III., English seaborne trade obtained a firm footing in Italy and other Mediterranean countries.

We, fortunately, possess drawings which show that an enormous advance was made in shipbuilding during the period under discussion, or that, at any rate, the advance had by that time reached England. Fig. 37 illustrates a large ship of the latter half of the fifteenth century. It is taken from a manuscript in the Cottonian Library, by John Rous, the celebrated Warwickshire antiquary and historian. This manuscript records the life and history of Richard Beauchamp, Earl of Warwick, who was born in 1381, and died in 1439. The author of the manuscript, however, lived till 1491, in the early part of the reign of Henry VII., and we may therefore conclude that the illustrations represent ships of the latter half of the fifteenth century. The vessel shown in Fig. 37 was used for war purposes, as four guns were mounted on the broadside. There were also four masts and a bowsprit, and a strongly developed forecastle, which formed part of the structure of the ship. There was apparently very luxurious accommodation provided for passengers and officers in a large deck-house at the poop. The mainsail was of very large dimensions, and86 was emblazoned with the arms of the Earl of Warwick. In this illustration we see an early approach to the modern type of sailing-ship. There are several other drawings of ships in the same manuscripts, and most of them have the same general characteristics as Fig. 37.

Fig. 37.—English ship. Latter half of fifteenth century.

The reign of Henry VII. (1485 to 1509) was a memorable one in the annals of navigation and commerce. Two years87 after he came to the throne, the Portuguese sent the expedition, previously referred to, to discover a route to the Indies round Africa. The expedition never reached its destination, but Diaz succeeded in discovering the Cape of Good Hope.

Fig. 38.—Columbus' ship, the *Santa Maria*, 1492.

A few years later, in 1492, Christopher Columbus made his famous attempt to reach the Indies by sailing west. This expedition, as is well known, resulted in the discovery of the West Indian Islands, and, shortly afterwards, of the mainland of America. The ships which Columbus took with him on his voyage were three in number, and small in size. As Spain had possessed many large vessels for a century and a half before the time of Columbus, it is probable that he was entrusted with small ships only, because the Government did not care to risk much capital in so adventuresome an undertaking.

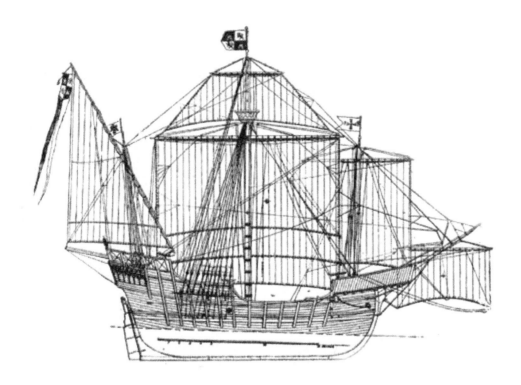

Fig. 39.—Sail-plan of the *Santa Maria*.

Fortunately, we have a fairly exact knowledge of the form and dimensions of the caravel *Santa Maria*, which was the largest of the three vessels. She was reconstructed in 1892-9389 at the arsenal of Carraca, by Spanish workmen, under the superintendence of Señor Leopold Wilke, for the Chicago Exhibition of 1893. Señor Wilke had access to every known source of information. Figs. 38 to 40 give a general view, sail-plan and lines, of this ship as reconstructed.

The following were her leading dimensions:—

Length of keel	60·68	feet
Length between perpendiculars	74·12	"
Extreme length of ship proper	93	"
Length over all	128·25	"
Breadth, extreme	25·71	"
Displacement fully laden	233	tons
Weight of hull	90·5	"

The *Santa Maria*, like most vessels of her time, was provided with an extensive forecastle, which overhung the stem nearly 12 ft. She had also an enormous structure aft, consisting of half and quarter decks above the main deck. She had three masts and a bowsprit. The latter and the fore and main masts were square-rigged, and the mizzen was lateen-rigged. The outside of the hull was strengthened with vertical and longitudinal timber beams.

The *Santa Maria*, as reproduced, was sailed across the Atlantic from Spain by Captain D. V. Concas and a Spanish crew in the year 1893. The course taken was exactly the same as that followed by Columbus on his first voyage. The time occupied was thirty-six days, and the maximum speed attained was about 6½ knots. The vessel pitched horribly.

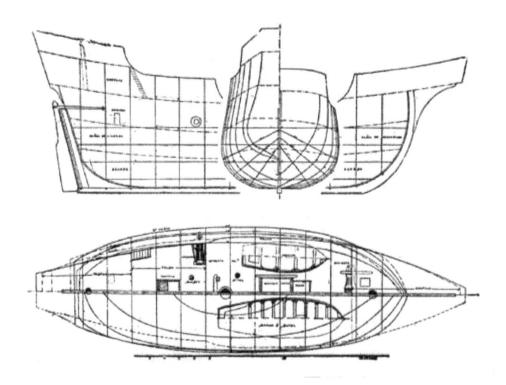

Fig. 40.—Lines of the *Santa Maria*.

In 1497 the first English expedition was made to America under John Cabot. We have no particulars of the ship in which Cabot sailed, but it could not have been a large one, as it is known that the crew only numbered eighteen. The expedition sailed from Bristol in the month of May, and land, which was probably Cape Breton, was sighted on June 24. Bristol was reached on the return journey at the end of July. In the following year Cabot made another voyage, and explored the coast of North America from Cape Breton to as far south as Cape Hatteras. Many other expeditions in the same direction91 were fitted out in the last years of the fifteenth and the first years of the sixteenth centuries.

While Cabot was returning from his first voyage to North America, one of the most famous and most epoch-making expeditions of discovery of modern times was fitted out in Portugal. On July 24, 1497, Vasco da Gama set sail from the Tagus in the hope of reaching India *via* the Cape of Good Hope. His squadron consisted of three ships, named the *San Gabriel*, the *San Raphael*, and the *Birrio*, together with a transport to carry stores. There is a painting in existence at Lisbon of the *San Gabriel*, which is supposed to be authentic. It represents her as having a high poop and forecastle, very like the caravel *Santa Maria*. She had four masts and a bowsprit. The latter and the fore and main masts were square-rigged. The *San Gabriel* was, however, a much larger vessel than the *Santa Maria*. She is said to have been constructed to carry 400 pipes of wine. This would be equivalent to about 400 tons measurement, or, from 250 to 300 tons register.15 The other two ships selected were of about the same dimensions, and of similar equipment and rig, in order that, in the event of losses, or accidents, each of the ships might make use of any of the spars, tackle, or fittings belonging to the others.

It may here be mentioned that the ships reached Quilimane, on the east coast of South Africa, on January 22, 1498. After many visits to East African ports, during which they satisfied themselves that the arts of navigation were as well understood by the Eastern seamen as by themselves, they set sail for India early in August, and after a voyage of twenty, or, as some say, twenty-three days, they sighted the coast, and shortly afterwards arrived in Calicut, nearly fourteen months after they started from Lisbon.

92 About this time the Memlook Sultans of Egypt absolutely cut off the trade which had been carried on for centuries between the Italian Republics and the Malabar coast of India *via* the overland route and the Red Sea. It was this fact that gave the discovery of the sea-route to India such enormous importance, and, ultimately, it was one of the causes of the commercial downfall of the Italian Republics. The Cape route became the great high-road of commerce to the East, and remained so down to the present reign, when the re-establishment of the overland route, and, eventually, the successful cutting of the Suez Canal, restored commerce to its old paths.

The discoveries of Columbus, Vasco da Gama, John Cabot, and their successors, had an enormous influence upon shipbuilding, as they not only widened the area of seaborne commerce, but offered strong inducements to navigators to venture on the great oceans, far from land, in craft specially adapted for such voyages. Hitherto, sailors had either navigated the great inland seas of Europe or had engaged in the coasting trade, and the longest voyages undertaken before the end of the fifteenth century were probably those which English merchants made between Bristol and Iceland, and between our Eastern ports and Bergen.

Henry VII. not only encouraged commerce and voyages of discovery, but also paid great attention to the needs of the Royal Navy. He added two warships to his fleet, which were more powerful vessels than any previously employed in this country. One of them, named the *Regent*, was copied from a French ship of 600 tons, and was built on the Rother about 1490. She carried four masts and a bowsprit, and was armed with 225 small guns, called serpentines. The second ship was named the *Sovereign*, and it is remarkable, as showing the connection at that time between land and naval architecture, that she was built under the superintendence of Sir Reginald93 Bray, who was also the architect of Henry VII.'s Chapel at Westminster Abbey, and of St. George's Chapel, Windsor. The *Sovereign* carried 141 serpentines.

The *Regent* was burnt in an action off Brest in the reign of Henry VIII., in the year 1512. She caught fire from a large French carrack, called the *Marie la Cordelière*, which she was attacking. Both ships were utterly destroyed.

The *Marie la Cordelière* was probably the largest warship of her time. She is said to have carried 1,200 men, and to have lost 900 killed in the action. She was built at Morlaix at the sole cost of Anne of Brittany, then Queen of France.

Fig. 41.—The *Henry Grace à Dieu. Pepysian Library, Cambridge.*

The *Regent* was replaced by a very famous ship called the *Henry Grace à Dieu*, otherwise known as the *Great Harry*. As a consequence, most probably, of the size and force of some of the French ships, as revealed in the action off Brest, the *Henry Grace à Dieu* was a great advance on any previous British94 warship. She was built at Erith, and was probably launched in June, 1514. Her tonnage is given in a manuscript in Pepys' "Miscellanies" as 1,500; but it is generally believed that she did not in reality exceed 1,000 tons.

Fig. 42.—The *Henry Grace à Dieu. After Allen.*

There are more drawings than one in existence, supposed to represent this famous warship. One of them, shown in Fig. 41, is from a drawing in the Pepysian Library, in Magdalene College, Cambridge. Another, shown in Fig. 42, is from an engraving by Allen of a picture ascribed to Holbein. The two illustrations differ in many important respects and cannot both represent the same ship. There is very little doubt that Fig. 41 is the more correct representation of the two, because it is confirmed in all essential respects by Volpe's picture of the embarkation of Henry VIII. at Dover in 1520 on this very ship. Volpe's picture is now at Hampton Court Palace, and shows four other ships of the Royal Navy, which were all built in the same style as the Pepysian drawing of95 Fig. 41, with enormous forecastles and poops. The vessel represented in the picture ascribed to Holbein appears to belong to a later date than 1520, and is, in fact, transitional between the ships of this period and those of the reign of Elizabeth. One of the warships of the latter period is shown in Fig. 45.

According to a manuscript, in the Pepysian Collection, the *Henry Grace à Dieu* was armed with twenty-one guns and a multitude of smaller pieces. The numbers of the various guns and the weights of their shot are given in the following table:—

Name of gun.	Number.	Weight of shot.
		lbs.
Cannon	4	60
Demi-cannon	3	32
Culverin	4	18
Demi-culverin	2	8
Saker	4	6
Cannon Perer	2	26
Falcon	2	2

The sizes of the guns of this time are pretty accurately known, because one of the ships of Henry VIII., called the *Mary Rose*, built in 1509, went down off Portsmouth in 1545, and several of her guns have been recovered, and are still in existence.

The por-holes were circular, and so small in diameter that no traverse could have been given to the guns. This practice continued to prevail till the time of the Commonwealth. There were five masts in this, as in all other first-rates henceforth down to the time of Charles I. One of the masts was inclined forward, like a modern bowsprit. Each mast was made in one piece, the introduction of separate topmasts having been a more modern improvement.

Fig. 43—Genoese carrack. 1542.

The highest development in the art of shipbuilding at this96 period was reached in the large merchant-ships called Carracks. The competition between the great trading republics of Italy, viz. Venice and Genoa, and the rivalry of Portugal probably accounted for the marked improvement in the character of merchant-ships in the fifteenth and sixteenth centuries. Fig. 43 gives a representation of a large Genoese carrack of the sixteenth century. It will be noticed that this vessel had four masts, and was square-rigged, the foremost mast having been inclined forward somewhat after the fashion of the modern bowsprit. In the sixteenth century the carrack often attained the size of 1,600 tons. Towards the latter half of this century a Portuguese carrack captured by the English was, in length, from the beakhead to the stern, 165 ft.; beam, 47 ft.; length of keel, 100 ft.; height of mainmast,97 121 ft.; circumference at partners, 11 ft.; length of mainyard, 106 ft.; burthen, 1,600 tons. This vessel carried 32 pieces of brass ordnance—a very necessary addition to the merchant-ship of the period—and accommodated between 600 and 700 passengers.

The most important maritime event in the sixteenth century was, undoubtedly, the fitting out by Spain, in 1588, of the gigantic expedition intended to invade this country in the reign of Queen Elizabeth. An account of the fleets on either side may therefore be interesting.

Fig. 44.—Spanish galleass. 1588.

The great Armada consisted of no less than 132 vessels, of which only four were galleys, and four galleasses.16 Of the remainder, 30 were under 100 tons, and 94 were between 130 and 1,550 tons. The total tonnage of the ships, less the galleys and galleasses, was 59,120. The armament consisted of 2,76117 guns. The seamen numbered 7,865 and the soldiers 20,671. The fleet was divided into ten squadrons. The largest vessel was the flagship of the Levant squadron, and was of 1,249 tons, and carried 30 guns. The crew consisted of 80 sailors and 98344 soldiers. The next largest was of 1,200 tons and carried 47 guns, but the greater number of the vessels were much smaller. The popular belief as to their incredible size and unwieldiness must therefore be dismissed as baseless, for even the largest ships were far exceeded in size by some of the carracks, or merchant vessels, of that day. On the average the Spanish vessels mounted 22 guns apiece, and carried crews of 231 sailors and soldiers. Fig. 44 is a sketch, taken from the tapestry of the old House of Lords, of one of the galleasses of the fleet. It will be noticed that she carried her guns extremely high, a peculiarity which was common to many of the Spanish vessels; for we read that their fire did more harm to the rigging than to the hulls of the English vessels.

The fleet mustered by Elizabeth was far more numerous, but its tonnage did not amount to one-half of that of the Armada. The total number of vessels sailing under the English flag was 197, of which, however, only 34 belonged to the Royal Navy. The remainder were merchant vessels, hastily fitted out and adapted for purposes of war by their owners, or by the ports to which they belonged. Of the Royal ships the largest was the *Triumph*, built in 1561. She was commanded by Sir Martin Frobisher, and was only exceeded in size by four of the Spanish vessels. The *Triumph* was between 1,000 and 1,100 tons, but there were only seven ships in the English Navy of between 600 and 1,000 tons, whereas the Spaniards had no fewer than 45. The crew of the *Triumph* numbered 500, of whom 300 were sailors, 40 gunners, and 160 soldiers.

The *Triumph* carried 42 guns, of which 4 were cannon, 3 demi-cannon, 17 culverins, 8 demi-culverins, 6 sakers, and 4 small pieces. The greatest number of guns carried by any ship in the fleet was 56, mounted on board the *Elizabeth Jones*, of 900 tons, and built in

1559. The flagship of the Lord High Admiral, Lord Howard of Effingham, the *Ark*, was the most99 modern of the English warships, having been built in 1587. She was of 800 tons, carried a crew of 430, and mounted 55 guns.

Of the merchant auxiliaries the two largest were the *Galleon Leicester* and the *Merchant Royal*, each of 400 tons, and each carried a crew of 160 men. In the former of these the explorer Cavendish afterwards made his last voyage. Another of the merchant-ships, the *Edward Bonaventure*, belonged to the Levant Company, and in the years 1591 to 1593 was distinguished as the first English ship that made a successful voyage to India.

The size of a large number of the merchant-ships was under 100 tons. The total number of the crews of the entire English fleet was 15,551; of these 6,289 belonged to the queen's ships.

As a general rule, the English ships in the reign of Queen Elizabeth, both in the Royal Navy and in the Mercantile Marine, were much inferior in size to the vessels belonging to the great Maritime Republics of Italy and to Spain and Portugal. Hitherto the practice had been general of hiring Genoese and Venetian carracks for mercantile purposes. It is stated that about the year 1578, or twenty years after Queen Elizabeth's accession to the throne, there were only 24 ships in the Royal Navy and 135 of above 100 tons burthen in the whole kingdom, and but 656 that exceeded 40 tons. Nevertheless, in this reign there was a great development of mercantile activity, in which the sovereign as well as her people participated. Many trading expeditions were sent out to the West Indies and to North America, and warlike descents on the Spanish ports were frequently carried out, and were attended with great success. In Elizabeth's time the first British colony, Virginia, was founded in North America, and Sir Francis Drake undertook his memorable and eventful voyage round the world in a squadron, which consisted, at the100 commencement, of five vessels, whereof the largest, the *Pelican*, was of only 100 tons burthen, and the smallest a pinnace of 15 tons. So great was the progress made about this time in English maritime trade that, only four years after the date above mentioned, there were said to have been no less than 135 English commercial vessels of above 500 tons in existence.

In the year 1587 Drake, in his famous marauding expedition in the Spanish seas, captured a great carrack called the *San Felipe*, which was returning home from the East Indies. The papers found in her revealed the enormous profits which the Spaniards made out of their trade with India, and afforded such valuable information that the English merchant adventurers were incited to cut in and try to secure some share of this trade for themselves. This led, ultimately, to the founding of the celebrated East India Company, and to the conquest of India by the British. In 1589 certain merchants petitioned the queen to grant them a licence to trade with the East Indies; but Elizabeth, fearing the resentment of the Spanish and Portuguese, would not grant their request for many years, and it was not till the last day of the year 1599 that she gave a charter of incorporation to the Earl of Cumberland and 215 knights and merchants for fifteen years, and thus founded the first East India Company. English adventurers, however, did not wait for a charter before commencing their trading operations with the East, for in 1591 an expedition consisting of three ships was sent out under the command of James Lancaster. Only one of the three—

the *Edward Bonaventure*, which, as already mentioned, had been a merchant auxiliary in the English fleet that opposed the Armada—ever reached the East Indies in safety.

A few weeks after the charter had been granted Lancaster led another expedition to the East. His fleet consisted of101 five ships; the largest, the *Dragon*, was of 600 tons, and had a crew of 202. After an adventurous voyage the fleet returned to England in September, 1602, having been absent two years and eight months.

There is abundant evidence to show that foreign merchant ships in Elizabeth's reign were often much larger than any built in this country. The following are examples. In 1592 a Portuguese carrack called the *Madre de Dios* was captured and brought home. She was of 1,600 tons burthen, 165 feet long from stem to stern, and had seven decks, including the numerous half and quarter decks which formed the poop. In 1594 a Spanish carrack was destroyed which had 1,100 men on board. When Cadiz was taken in 1596 two Spanish galleons of 1,200 tons were captured, and the flagship, the *San Felipe*, of 1,500 tons, was blown up. In 1602 a Portuguese carrack of 1,600 tons was captured at Cezimbra. She was named the *San Valentino*, and was worth, with her cargo, a million ducats.

The system of striking topmasts appears to have been introduced into the English Navy in the reign of Queen Elizabeth. It is mentioned by Sir Walter Raleigh as a recent improvement and "a wonderful ease to great ships, both at sea and in the harbour." Amongst the other novelties mentioned by the same authority was the use of chain-pumps on board ship; they lifted twice the amount of water that the old-fashioned pumps could raise; studding, top-gallant, sprit and topsails were also introduced, and the weighing of anchors by means of the capstan. He also alludes to the recent use of long cables, and says that "by it we resist the malice of the greatest winds that can blow." The early men-of-war, pierced with portholes, carried their lower guns very near the water. In some cases there were only fourteen inches from the lower sill of the portholes to the water-line. This practice led to many accidents; amongst others may be mentioned the loss of the102 *Mary Rose*, one of the largest ships in the Royal Navy in the time of Henry VIII. Sir Walter Raleigh mentions that, in his time, the practice was introduced of raising the lower tier of ports. Nevertheless, this improvement did not become general till the time of the restoration of Charles II. Fig. 45 is a representation of an English ship of war of the time of Queen Elizabeth, supposed to be of the date 1588. It is copied from the tapestries of the old House of Lords. It shows clearly the recently introduced topmasts alluded to by Sir Walter Raleigh. It is certainly a much more ship-shaped and serviceable craft than the vessels of Henry VIII. There is also in existence a drawing of a smaller Elizabethan warship in the Rawlinson MSS. in the Bodleian Library; in essential particulars, it confirms Fig. 45. Both of these show that the forecastles and poops had been considerably modified.

Fig. 45.—English man-of-war. About 1588.

Fig. 46.—Venetian galleass. 1571.

Another great naval war was waged in the latter half of the sixteenth century, about sixteen years before the defeat of the Spanish Armada. The scene was the Adriatic Sea, and the combatants were Venice, with her allies, Spain and the Papal States, on the one hand, and the Turks on the other. It culminated in the complete defeat of the latter at Lepanto in 1571. The site of the battle of Lepanto is very near to that of Actium, and it is a remarkable circumstance that twice in history a decisive naval battle between the West and East should have been decided at the same spot. The allies possessed a fleet consisting of 208 galleys and 6 galleasses. The Venetians introduced the latter type of vessel in order to meet the Turks on even terms. It was an improved form of galley with three masts, carrying several guns on the broadside, most of them mounted on the upper deck. Fig. 46 represents one of the Venetian galleasses as used at the battle of Lepanto, to the winning of which engagement they are said to have contributed materially. The galleass was essentially a Mediterranean warship. It was never generally adopted by the104 Western powers, but four Neapolitan vessels of this category, carrying each 50 guns, formed a part of the great Armada sent by Spain to effect the conquest of England. The galleass represented in Fig. 46 had a circular forecastle in which were mounted several guns, to be used in end-on attack.

It is impossible to read the accounts of the battle of Lepanto and of the defeat of the Spanish Armada without noticing the great contrast between the ships used in the two wars at about the same period. In the Mediterranean the single-banked galley was still the prevailing type, while in the Western and Northern seas the bulk of the Spanish and the whole of the British fleets were sailing-ships.

It does not appear that any further novelties, or improvements, worth alluding to were introduced into the practice of shipbuilding till the accession of the House of Stuart in 1603. All the monarchs of this family paid particular attention to the development of the Royal Navy. King James I. had in his service an educated naval architect of the name of Phineas Pett, who was a Master of Arts of Emmanuel College, Cambridge, and a member of a famous family of shipbuilders who had been employed for two centuries previously, from father to son, as officers and architects in the Royal Navy. Some time after the accession of James, a Royal Commission inquired into the general state and management of the navy, and issued a report in 1618, which was in effect "a project for contracting the charge of His Majesty's Navy, keeping the coast of England and Ireland safely guarded, and his Majesty's ships in harbour as sufficiently guarded as now they are, provided that the old debts be paid, ... and certain assignments settled for the further payment of the navy quarterly." At the time the report was issued there were only seventeen vessels in the navy which had been built during the reign of James. The most important of these was the *Prince Royal,* built in 1610,105 106and, at the time, considered to be one of the finest men-of-war in the world. Fig. 47 is an illustration of a man-of-war of the period, which, there is strong evidence for believing, was this very vessel. It was designed and built under the superintendence of Phineas Pett at Woolwich Dockyard, and was given by the king to his son Henry, Prince of Wales, in honour of whom it was named the *Prince Royal.* It was in many respects a remarkable departure from the prevailing practice of the times, and, if stripped of its profuse carved work, was very similar in outline to the men-of-war built as recently as the commencement of the last century. The designer was bold enough to abandon some of the time-honoured features of ship construction, such as the beak, or prow, derived from the old galleys, and the square buttock, or tuck. The latter feature, however, continued to appear in the ships of most other European countries for some time afterwards. The length of keel of this vessel was 114 ft., and the beam 44 ft. The reputed burthen was 1,400 tons, and the vessel was pierced for 64 guns, whereof she carried 55, the vacant portholes being filled in action from the opposite side, a custom which prevailed down to the last century and was adopted in order to lessen the dead weight carried aft. The great difference between the shape of the quarter galleries and forecastle in this ship and in the earlier types will be noted. The armament of the *Prince Royal* consisted of the following guns: On the lower deck six 32-pounders, two 24-pounders, and twelve 18-pounders. The bow and aftermost ports were empty, and in case of necessity the former was filled by an 18-pounder from the opposite side, and the latter by a 24-pounder from the stern-ports. The upper deck was armed with 9-pounders, the aftermost port being vacant, and filled up when required. The quarter-deck and forecastle were provided with 5-pounders.

Fig. 47.—The *Prince Royal.* 1610.

107The building of this ship aroused many apprehensions, and a Commission was appointed to report on the design while it was being constructed. It certainly seems that gross errors were made in the calculations. For instance, it was estimated that 775 loads of timber would be required for her construction, whereas 1,627 loads were actually used. The timber also was so unseasoned that the ship only lasted fifteen years, and had then to be rebuilt.

Many complaints were made about this time of the incapacity and ignorance of English shipbuilders. Sir Walter Raleigh laid down the following as the principal requirements of warships: strong build, speed, stout scantling, ability to fight the guns in all weathers, ability to lie to easily in a gale, and ability to stay well. He stated that in all these qualities the royal ships were deficient. He also called attention to the inferiority of our merchant-ships, and pointed out that, whereas an English ship of 100 tons required a crew of thirty hands, a Dutch vessel of the same size would sail with one-third of that number.

Another authority of the time complained that—

"he could never see two ships builded of the like proportion by the best and most skilful shipwrights ... because they trust rather to their judgment than their art, and to their eye than their scale and compass."

The merchant navy of England languished during the early years of the reign of James I. Owing, however, to the patronage and assistance extended by the king to the East India

Company, and also in no small measure to the stimulus caused by the arrival of some large Dutch merchantmen in the Thames, the merchants of London abandond the practice of hiring ships from foreigners and took to building for themselves. In the year 1615 there were not more than ten ships belonging to the Port of London with a burthen in excess of 200 tons, but, owing to the sudden development of shipbuilding, the Port108 of Newcastle in the year 1622 owned more than 100 ships exceeding the above-mentioned tonnage.

In the year 1609 the king granted a new charter to the East India Company, and in the following year a vessel, called the *Trade's Increase*, was sent out. This ship was the largest merchantman built up to that time in England. Her career, however, was not fortunate. She was careened at Bantam, in order that some repairs to her hull might be effected, but she fell over on her side and was burnt by the Javanese.

Before the year 1613 British merchants had made altogether twelve voyages to the East Indies, for the most part in ships of less than 500 tons. In that year, however, all the merchants interested in the Oriental trade joined together to form the United East India Company. The first fleet fitted out by the re-organised Company consisted of four ships, of 650, 500, 300, and 200 tons burthen respectively. It had to fight its way with the Portuguese before it could commence to trade. The Portuguese considered that they were entitled to a monopoly of the trade with the East, and jealously resented the intrusion of the English merchantmen, whom they attacked with a fleet of six galleons, three ships, two galleys, and sixty smaller vessels. They were, however, ignominiously defeated, and the English merchants were enabled to accomplish their purpose.

During the last five years of the reign of James I. the strength of the Royal Navy was increased twenty-five per cent. His son and successor, Charles I., through all the troubles of his eventful reign, never neglected this branch of the national defences, and during his reign the Mercantile Marine grew to such an extent that, at the time of the outbreak of the Civil War, the port of London alone was able to furnish 100 ships of considerable size, all mounting cannon and fitted up in every respect for the operations of war.

Fig. 48.—The *Sovereign of the Seas*. 1637.

110The *Sovereign of the Seas*, illustrated in Fig. 48, may be taken as a sample of the largest type of warship built by Charles. Like the *Prince Royal*, she was designed by Pett, and was considered to be the most powerful man-of-war in Europe of her time. Her construction must have been a great improvement on that of the *Prince Royal*; for, whereas the latter ship was declared to be no longer fit for service fifteen years after her launch, the *Sovereign of the Seas*, though engaged in most of the naval battles of the seventeenth century, remained in good condition for a period of sixty years, and was then accidentally burnt at Chatham when about to be rebuilt. She was the first three-decker in the Royal Navy, but as she proved somewhat crank, she was cut down to a two-decker in the year 1652. At the Restoration she was renamed the *Royal Sovereign*.

This very remarkable vessel was of 1,683 tons burthen. Her length of keel was 128 ft.; length over all, 167 ft.; beam, 48 ft. 4 in.; and depth from top of lanthorn to bottom of keel, 76 ft. She was built with three closed decks, a forecastle, a half-deck, a quarter-deck, and a round-house. She carried in all 102 or 104 guns, and was pierced for thirty guns on the lower, thirty on the main, and twenty-six on the upper deck; the forecastle had twelve, and the half-deck fourteen ports. She also carried ten chasers forward, and as many aft. She was provided with eleven anchors, of which one weighed two tons.

The *Royal Sovereign* may fairly be taken as representing the commencement of a better school of ship construction. Her merits were due to the talents of Phineas Pett, who, though not uniformly successful in his earlier designs, was a great innovator, and is generally regarded as the father of the modern school of wooden shipbuilding.

Very little is known, unfortunately, of the character and rig111 of the smaller classes of trading vessels of the end of the sixteenth and the commencement of the seventeenth centuries. It is, however, tolerably certain that cutter-rigged craft were used in the coasting and Irish trades as far back as 1567; for there is a map of Ireland of that date in existence on which are shown two vessels rigged in this manner.

With the description of the *Royal Sovereign* we close the account of mediæval naval architecture. Thanks to the fostering care of Charles I., to the genius of Pett, and to the great natural advantages conferred by the superiority of English oak to other European timbers, England at this period occupied a high place in the art of shipbuilding. The position thus gained was maintained and turned to the best advantage in the period of the Commonwealth, when successful naval wars were undertaken against the Dutch and other European States. These wars eventually resulted in establishing England, for a time, as the foremost maritime power in Europe.

CHAPTER V.

MODERN WOODEN SAILING-SHIPS.

The naval wars which followed the establishment of the Commonwealth contributed in a very large degree to the progress of shipbuilding. In 1652 war broke out with the United Provinces, headed by the Dutch, who were, prior to that period, the foremost naval and mercantile power in the world. The struggle lasted about two years, and during its continuance the British fleet increased from fifty-five first, second, and third rates, to eighty-eight vessels of corresponding classes, while a proportionately larger increase was made in ships of smaller denominations, and, in addition, the vessels lost in the war were replaced. The war with the Dutch was an exceptionally severe struggle, and ended in the complete victory of this country, which then stepped into Holland's place as foremost naval power. In addition to this war, Cromwell undertook an expedition to the Mediterranean, to punish the piratical states of Algiers, Tunis, and Tripoli. The fleet was commanded by Blake, and was completely successful in its operations, which resulted in a security for British commerce with the Levant that had never been known before. Admiral Penn was at the same time entrusted with the command of a powerful expedition to the Spanish West Indies. The annexation of Jamaica followed, and British commerce in the West increased. In fact, with the progress of the national navy the commerce of the country also extended itself, and the increased experience thus obtained in shipbuilding, both for the war and trading fleets, necessarily resulted in great improvements in the art.

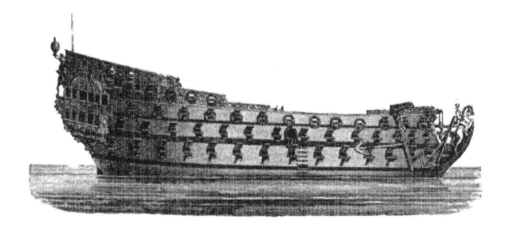

Fig. 49.—The *Royal Charles*. 1673.

114The expenditure on the navy in the time of the Commonwealth was enormous relatively to the total national revenue. In the year 1656-57 four-fifths of the income of the country was devoted to the sea service, in the following year two-thirds, and in 1658-59 nearly three-fifths. These are figures which have never been approached at any other period. The ships built during this time were of moderate dimensions. Only four were of 1,000 tons. These were the *Dunbar*, of 1,047 tons and 64 guns, built in 1656; the *London*, built in the same year, of the same tonnage and number of guns, though of different dimensions; the *Richard*, of 1,108 tons and 70 guns, built in 1658; and the *Naseby*, built in 1655, of 1,229 tons and 80 guns. All four were renamed at the Restoration.

Charles II. and his brother, the Duke of York, afterwards James II., both possessed in an eminent degree the fondness for the navy which distinguished all the members of the Stuart dynasty, though, unfortunately, after the first naval war waged by Charles against Holland, the condition of the fleet was allowed to deteriorate very rapidly. As a sample of the type of warship of the first class built in this reign, we give, in Fig. 49, the *Royal Charles*, which was constructed at Portsmouth dockyard in 1673, by Sir Anthony Deane, to carry 100 guns. This illustration and that of the *Sovereign of the Seas* are after pictures by Vandevelde. This ship was the largest in the navy, excepting always the famous old *Sovereign of the Seas* and the *Britannia*. The latter was built at Chatham, by Pett, in 1682, and carried 100 guns, and measured 1,739 tons. The *Royal Charles* created as much sensation in its day as did the famous ship built for Charles I. There is a beautiful model of the *Royal Charles* in the Museum.

Fig. 50.—The *Soleil Royal*. 1683.

The following table gives the leading dimensions of the *Royal Charles* and the *Britannia*:—

Name of ship.	Length.	Breadth.		Depth of hold.		Draught.		Complement.
	ft.	ft.	in.	ft.	in.	ft.	in.	
Royal Charles	136	46	0	18	3	20	6	780
Britannia	146	47	4	19	7½	20	0	780

Fig. 50 is an illustration after Vandevelde of a famous French first-rate of the same period, named the *Soleil Royal*, of 106 guns. She was destroyed in Cherbourg Bay the day after the battle of Cape La Hogue, in 1692. Fig. 51 is a Dutch first-rate, named the *Hollandia*, of 74 guns. She was built in 1683, and took part in the battle of Beachy Head as flagship of Admiral Cornelis Evertsen.

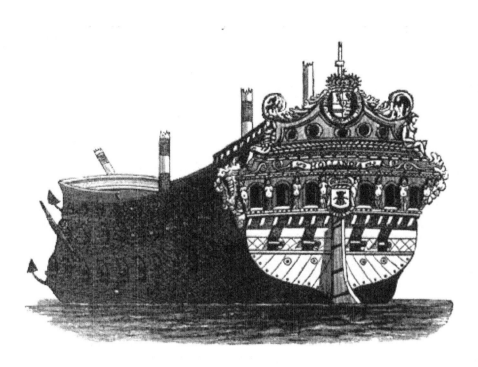

Fig. 51.—The *Hollandia*. 1683.

The chief difference between the British and foreign builds of warship of the latter half of the seventeenth century was that the English vessels were always constructed with the rounded117 tuck before mentioned, as introduced by Pett, while the Continental ships all had the old-fashioned square tuck, which is well illustrated in Fig. 51. The Dutch ships in one respect excelled all others, in that they were the first in which the absurd practice of an exaggerated "tumble home," or contraction of the upper deck, was abandoned. This fashion was still carried out to a very great extent by the English, and to a less extent by the French and Spaniards. The chain-plates in the English vessels were also fixed extremely low, while the Dutch fixed them as high as the sills of the upper-deck ports would allow. In consequence of the shallowness of the Dutch harbours, the draught of their ships was also considerably less than that of the English vessels of corresponding force.

Most of the ships in a seventeenth-century fleet deemed fit to take their station in the line of battle were third-rates. The first and second rates were exceptional vessels, and were only employed in particular services. A comparative table of the dimensions and armament of the various rates, or classes in the year 1688, is annexed:—

Designation.	Length of keel.	Breadth.	Depth of hold.	Draught of water.	Tons.	Guns on war service at home.	Crew.
	Feet.	Feet.	Feet.	Feet.			
1st Rate	128 to 146	40 to 48	17.9 to 19.8	20 to 23.6	1100 to 1740	90 to 100	600 to 815
2nd Rate	121 to 143	37 to 45	17 to 19.8	16 to 21	1000 to 1500	82 to 90	540 to 660
3rd Rate	115 to 140	34 to 40	14.2 to 18.3	16 to 18.8	750 to 1174	60 to 74	350 to 470
4th Rate	88 to 108	27 to 34	11.2 to 15.6	12.8 to 17.8	342 to 680	32 to 50	180 to 230
5th Rate	72 to 81	23.6 to 27	9.9 to 11	11.6 to 13.2	211 to 333	26 to 30	125 to 135

The first so-called frigate was designed by Peter Pett, and built at Chatham in 1646. She was named the *Constant Warwick*. Her dimensions were: length of keel, 85 ft.;118 breadth, 26 ft. 5 in.; depth, 13 ft. 2 in.; tonnage, 315; guns, 32; crew, 140. She worked havoc amongst the privateers of the time.

The bomb-ketch was originally introduced by a famous French naval architect named Bernard Renan, about 1679. This class of warship was first employed by Louis XIV. in the bombardment of Algiers, where it produced an enormous effect. Bomb-ketches were of about 200 tons burthen, very broad in proportion to their length, and built with great regard to strength, on account of the decks having to bear the downward recoil of the mortars. The latter were placed in the fore-part of the vessel, which was purposely left unencumbered with rigging. The hold between the mortars and keel was closely packed with old cables, cut into lengths. The yielding elastic qualities of the packing assisted in taking up the force of the recoil. The bombs weighed about 200 pounds, and the consternation and terror produced by them may readily be realized when it is remembered that, up to that time, the most dangerous projectile which a warship could discharge at a land fortification was a thirty-two pound shot. These vessels were fitted with two masts, one in the middle and the other in the stern.

While referring to this invention of Bernard Renan, it should be mentioned that France rose to the rank of a great naval power in the reign of Louis XIV., under the famous minister Colbert, in the latter half of the seventeenth century. When Louis succeeded to the throne the French Navy was practically non-existent, as it consisted only of four, or five, frigates. In 1672 he had raised the strength of the fleet to fifty line-of-battle ships and a corresponding number of frigates and smaller vessels. Nine years afterwards, the French marine numbered 179 vessels of all classes, exclusive of galleys. In 1690 the French fleet in the Channel alone numbered sixty-120119eight ships, while the combined British and Dutch squadrons consisted only of fifty-six, and suffered a defeat at Beachy Head, in which the English lost one vessel and their allies six. This defeat was, however, amply revenged two years afterwards, when the allies succeeded in opposing the enormous number of ninety-nine ships of the line, besides thirty-eight frigates and fireships, to Tourville's fleet of forty-four ships of the line and thirteen smaller vessels, and defeated it off Cape La Hogue, inflicting on it a loss of fifteen line-of-battle ships, including the famous *Soleil Royal*, of 108 guns, illustrated in Fig. 50. From the time of Louis XIV. down to the present date French naval architects have always exercised a most important influence on the design of warships, a circumstance which was largely due to the manner in which Colbert encouraged the application of science to this branch of construction.121 It may be truly said that, during the whole of the eighteenth century, the majority of the improvements introduced in the forms and proportions of vessels of the Royal Navy were copied from French prizes.

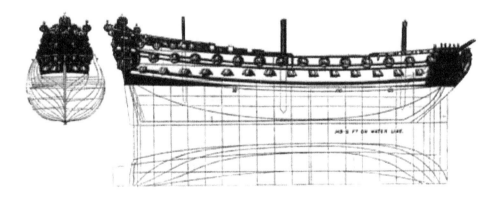

Fig. 52.—British second-rate. 1665.

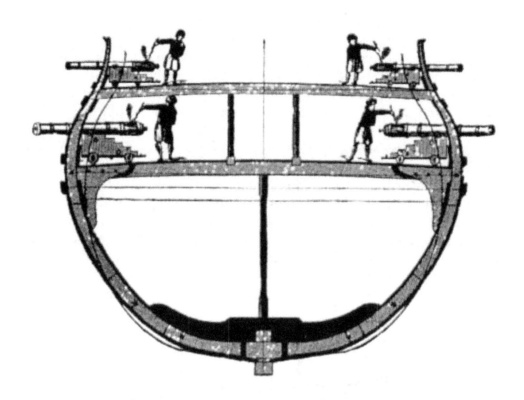

Fig. 53.—Midship section of a fourth-rate.

In order to complete the illustrations of British warships of the latter half of the seventeenth century views of a second-rate are given in Fig. 52, and a cross-section of a fourth-rate in Fig. 53.

It would be impossible in the present work to notice in detail all the alterations in size and structure of ships which took place during the eighteenth century. A few of the leading changes may, however, be mentioned. In the year 1706 an attempt was made to systematize the dimensions of the various rates, and the figures as given in the following table were fixed:—

Number of guns	90	80	70	60	50	40
Length of gun-deck	162 ft.	156 ft.	150 ft.	144 ft.	130 ft.	118 ft.
Extreme breadth	47 ft.	43 ft.	41 ft.	38 ft.	35 ft.	32 ft.
Depth of hold	18 ft. 6 in.	17 ft. 8 in.	17 ft. 4 in.	15 ft. 8 in.	14 ft.	13 ft. 6 in.
Tonnage	1552	1283	1069	914	705	532

When the figures were compared with those of contemporary French ships of the same rates, it was found that the British vessels of every class were of inferior dimensions. Whenever British men-of-war were captured by the French, the number of their guns was reduced. It was universally admitted that the French ships were superior in sailing qualities;

so much so was this the case that, whenever a French squadron was chased, the English-built ships in it were the first to be overtaken. The subject of the superiority in size of the French ships was constantly coming to the front, and in 1719 a new122 establishment was made for the dimension of ships in our Royal Navy, according to the following scale:—

Number of guns.	90	80	70	60	50	40
Increase of length	2 ft.	2 ft.	1 ft.	0	4 ft.	6 ft.
Increase of breadth	2 in	6 in.	1 ft.	1 ft.	1 ft.	1 ft. 2 in.
Increase of tonnage	15	67	59	37	51	63

In addition to the increase in dimensions, much improvement was made in the same year in the interior arrangements, and in the preservation of the timber of which ships were constructed. Up till this period both thick stuff and planks were prepared by charring the inner surface while the outer surface was kept wet, and this process was continued till the plank was brought to a fit condition for bending to the shape it was required to take. In this year, however, the process of stoving was introduced. It consisted in placing the timber in wet sand and subjecting it to the action of heat for such time as was necessary in order to extract the residue of the sap and to bring it to a condition of suppleness. In the year 1726 the process was favourably reported on by two of the master shipwrights in their report on the state of the planking on the bottom of the *Falkland*. Some of the planking had been charred by the old process, some stoved by the new, and the remainder had been neither stoved nor charred. The stoved planks were found to be in a good state of preservation, while many of the others were rotten. The process remained in use till 1736, when it was superseded by the practice of steaming the timber. The steaming and the kindred process of boiling remained in vogue during the whole of the remainder of the era of wooden shipbuilding. In 1771 the rapid decay of ships in the Royal Navy once more caused serious attention to be paid to the subject of the preservation of timber. It was, in consequence, arranged that larger stocks123 of timber should be kept in the dockyards, and that line-of-battle ships should stand in frame for at least a year, in order to season before the planking was put on. Similarly, frigates were to stand in frame for at least six months, and all thick stuff and planking was to be sawn out a year before it was used and stacked, with battens between the planks, so as to allow of the free circulation of the air. Similar regulations were put in force for the beam pieces, knees, and other portions of the ships.

Much trouble was caused by the injurious effects of bilge-water and foul air in the holds of ships, and various remedies were devised from time to time. In 1715 structural improvements were devised to allow of the bilge-water flowing more freely to the pumps, and trunks were fitted to the lower decks to convey air to the holds. In 1719 it was proposed that the holds of ships should have several feet of water run into them in the early spring in order to cool them, and that it should not be pumped out till August; but this remedy was never extensively practised. In 1753 Dr. S. Hales proposed a system of ventilation by means of windmills and hand-pumps, which produced excellent results. It was noticed that the accumulation of carbonic acid gas and foul damp air in the holds, not only set up rapid decay in the ship, but also most injuriously affected the health of the crews. Dr. Hales' system was employed in the *Prince* from 1753 to 1798, and it was considered that the durability of this vessel had been greatly increased. It was also reported by Lord Halifax that the mortality on the non-ventilated ships on the coast of Nova Scotia was twelve times as great as on those vessels which were fitted with Dr. Hales' appliances.

There are not many records in existence of the merchant-vessels of this period. Fig. 54 is a representation of an armed East Indiaman which was launched at Blackwall in 1752. Her length of keel was 108 ft. 9 in.; breadth, 34 ft.; and burthen, 124 125668 tons. She was named the *Falmouth*, and was constructed by the famous shipbuilder, John Perry, of Blackwall Yard. She was commenced almost exactly two years before the date of her launch. Like all her class, she was heavily armed.

Fig. 54.—The *Falmouth*. East Indiaman. Launched 1752.

At the close of the war against France and Spain, which lasted from 1744 to 1748, great complaints were made of the weakness of our warships at sea. It was also found that the establishment of 1719 had not been adhered to, and the dimensions of ships were not fixed in accordance with any particular standard. The first defect was remedied by the placing of as many standards of wood, or iron, on the different decks as could be conveniently arranged, so as not to interfere with the guns, and by the use of larger bolts than had hitherto been employed, as high up as possible in the throats of the hanging knees. Also the beams of the quarter-deck and round-house were supported with lodging knees, and in some instances with hanging knees of wood, or iron. Various other pieces, such as the stem, were also strengthened and the weights of the taffrails and quarter-pieces were reduced. The advice of the master shipwrights of the various dockyards was sought, in order to fix a new establishment of dimensions, but great difficulties were found in introducing the much-needed reforms, and for some time afterwards the ships of the British Navy were at a disadvantage with those of foreign countries by reason of their contracted dimensions and inferior forms.

The capture, with great difficulty, of a Spanish ship of seventy guns, named the *Princessa*, in 1740, by three British men-of-war of equal rating, but far inferior dimensions, was one of the events that first opened the eyes of the Admiralty to the defects of their vessels. The first attempt towards introducing a better type of ship was made in 1746, when the *Royal George*, famous for her size, her services, her beauty and126 misfortunes, was laid down. She was not launched till 1756. The following were her principal dimensions:—

Length of keel for tonnage	143 ft. 5½ in.
Length of gun-deck	178 ft.
Extreme breadth	51 ft. 9½ in.
Depth of hold	21 ft. 6 in.
Tonnage	2047
Number of guns	100
Crew	750 men.

Fig. 55 is an illustration of this ship. She rendered great services to the country under the orders of Admiral Lord Hawke, especially in the memorable defeat of the French Navy off the island of Belle-isle in 1759. She was lost at Spithead in 1782, when being inclined in order to have some repairs to her bottom executed. She capsized, and went under, 900 men, women, and children being drowned in her.

The *Royal George* was followed by several others of various rates and improved dimensions, notably by the *Blenheim* (90) and the *Princess Amelia* (80). The latter was one of the most famous ships of her day, and was constantly employed as long as she continued fit for service. In 1747 a French ship of seventy-four guns named the *Invincible* was captured, and was found to be such an excellent vessel that her dimensions were adopted for the *Thunderer*, laid down about 1758. One of the most interesting models in the Museum is of the *Triumph* (74), also built on the lines of the *Invincible* in 1764. Her length of gun-decks was 171 ft. 3 in.; breadth, 49 ft. 9 in.; depth of hold, 21 ft. 3 in.

In the following year was built the *Victory*, 100 guns, famous as Nelson's flagship at Trafalgar, and still afloat in Portsmouth Harbour. Her dimensions are: length of gun-deck, 186 ft.; breadth, 52 ft.; depth of hold, 21 ft. 6 in.; tonnage, 2,162.127

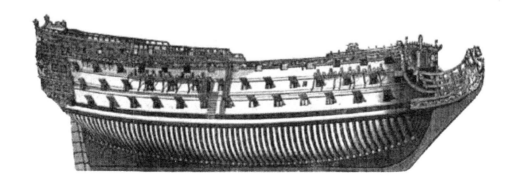

Fig. 55.—The *Royal George*. 1746.

The following table gives the dimensions of typical ships of war constructed about the middle of the eighteenth century:—

Number of guns	100	90	80	74	64	50
Length of gun-deck	178 ft.	176 ft. 1 in.	165 ft.	171 ft. 3 in.	159 ft. 4 in.	146 ft.
Length of keel for tonnage	143 ft. 6 in.	142 ft. 7 in.	133 ft.	138 ft. 8 in.	130 ft. 9½ in.	120 ft. 8½ in.
Extreme breadth	51 ft. 9½ in.	49 ft. 1 in.	47 ft. 3 in.	49 ft. 9 in.	44 ft. 6½ in.	40 ft. 4½ in.
Depth of hold	21 ft. 6 in.	21 ft.	20 ft.	21 ft. 3 in.	18 ft. 9½ in.	17 ft. 2 in.
Tonnage	2,047	1,827	1,580	1,825	1,380	1,046

The genuine frigate—that is to say, a large cruiser, of relatively high speed, carrying its main armament on one deck—was introduced into the Royal Navy in 1741, when the *Adventure* was built. She carried thirty-two guns, of which twenty-two were 12-pounders. The first British 36-gun frigates were the *Brilliant* and *Pallas*, built in 1757. Their main armament also consisted of 12-pounders. French frigates of the same date were of larger dimensions, as is proved by the following table which compares the principal measurements of the *Brilliant* and of the French frigate *Aurore*:—

Name of ship	Length of gun-deck		Breadth.		Depth of hold		Tonnage.	Complement.
	ft.	in.	ft.	in.	ft.	in.		
Brilliant	128	4	35	8	12	4	718	240
Aurore	144	0	38	8½	15	2	946	250

In the year 1761 a most important improvement was introduced, which greatly increased the usefulness of ships. This was the discovery of the value of copper plates as a material for sheathing their bottoms. Previously to this period lead was the metal used for sheathing purposes, and even it was only employed occasionally. In other cases the bottoms of vessels were paid over with various compositions, the majority of which fouled rapidly. The first vessel in the navy that was copper-sheathed was the *Alarm*, a 32-gun frigate. At first129 the use of copper caused serious oxidation of the iron bolts employed in the bottom fastenings, and copper bolts were substituted for them.

About the year 1788 the dimensions of the various rates were again increased in order to keep pace with the improved French and Spanish ships. In the year 1780 the 38-gun frigate founded on a French model was introduced into the navy, and continued to be much used throughout the great wars at the close of the eighteenth and the commencement of the nineteenth century. The first British frigate of this rating was the *Minerva*, which measured 141 ft. in length of gun-deck; 38 ft. 10 in. width of beam; 13 ft. 9 in. depth of hold, and 940 tons—figures which were evidently based on those of the *Aurore*, captured in 1758 (see p. 128). In 1781 and 1782 two very large French frigates were captured. Their names were the *Artois* and *Aigle*, and they exceeded in size anything in this class that had yet been built. The length of gun-deck measured 158 ft.; width, 40 ft. 4 in.; depth of hold, 13 ft. 6 in.; tonnage, 1,152; they each carried 42 guns and 280 men.

Again, in 1790, the force of new ships of the various rates was much increased. The largest line-of-battle ship then built was the *Hibernia*, of 110 guns. She was the first of her class introduced into the navy. Her dimensions were as follows:—Length on gun-deck, 201 ft. 2 in.; extreme breadth, 53 ft. 1 in.; depth of hold, 22 ft. 4 in.; burthen in tons, 2,508. The armament consisted of thirty 32-pounders on the lower deck, thirty 24-pounders on the middle, and thirty-two 18-pounders on the upper decks, while eighteen 12-pounders were mounted on the forecastle and quarter-deck. It is worthy of remark that, for some time previously, the large line-of-battle ships carried 42-pounders on the lower deck, but it was found that the 32-pounders could be loaded much more quickly, and that a great advantage arose in consequence.

Fig. 56.—The *Commerce de Marseille*. Captured 1792.

In the year 1792 the first 40-gun frigate, the *Acasta*, was built. This type of vessel was intended to replace the old 44-gun two-decker. The *Acasta* measured 150 ft. on deck; 40 ft. 9½ in. extreme breadth; 14 ft. 3 in. depth of hold; with a burthen of 1,142 tons. Her armament consisted of thirty 18-pounders on the main deck, and ten 9-pounder long guns on quarter-deck and forecastle.

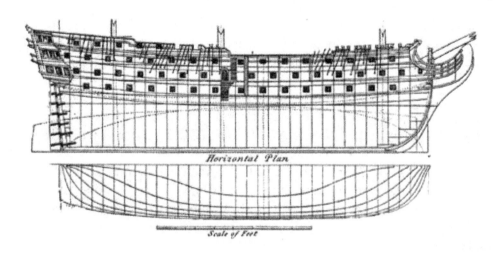

Fig. 57.—British first-rate. 1794.

During the whole of our naval history down to comparatively recent times, improvements in the dimensions and forms of our ships were only carried out after they had been originally adopted by the French, or Spaniards, or more recently by the people of the United States of America. Thus, we find that, shortly after war had been declared against the French Revolutionary Government in 1792, Admiral Hood took possession at Toulon, amongst other vessels, of a French first-rate called the *Commerce de Marseille*, which was larger and mounted more guns than any vessel in the service of Great Britain. Fig. 56 is an illustration of this fine man-of-war, which was 208 ft. 4 in. long on the lower deck, 54 ft. 9½ in. broad, of 25 ft. depth of hold, and of 2,747 tons burthen. As an instance of the progress in size, as related to armament, made during the century, we may compare the dimensions of this French first-rate with those of the *Royal Anne*, an English 100-gun ship built in 1706. The length of gun-deck of the latter ship was 171 ft. 9 in., and tonnage 1,809, the more recent vessel showing an increase of nearly fifty per cent. in tonnage for an increased armament of twenty guns.

As further examples of the naval architecture of this period, in Figs. 57 and 58 are given views of an English first-rate of the year 1794, and in Figs. 59 and 60 corresponding views of a heavy French frigate of about the year 1780.

One of the greatest improvements made at the end of the eighteenth century was the raising of the lower battery 133further above the water, so as to enable the heavy guns to be fought in all weathers. It was frequently observed that the old British men-of-war of seventy-four guns when engaging a hostile vessel to leeward were, on account of the crankness of the ship and the lowness of the battery, obliged to keep their lower ports closed; whereas the French ships, which were comparatively stiff, and carried their lower guns well above the water, were enabled to fight with the whole of their battery in all weathers.

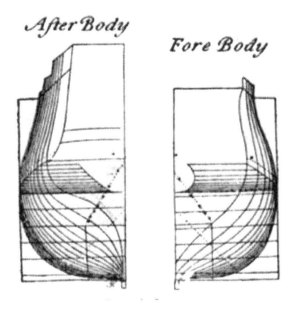

Fig. 58.—British first-rate. 1794.

After the capture of the *Commerce de Marseille*, an English first-rate, named the *Caledonia*, to carry 120 guns, was ordered to be laid down. She was not, however, commenced till 1805. Her dimensions and proportions closely approximated to those of her French prototype, and need not, therefore, be more particularly referred to. She was the first 120-gun ship built in this country.

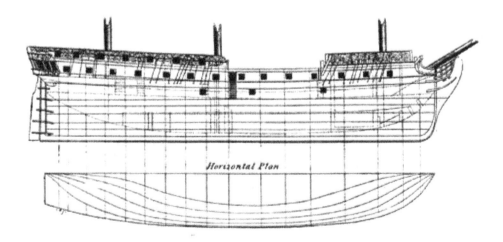

Fig. 59.—Heavy French frigate of 1780.

In the year 1812 the United States declared war against Great Britain. The struggle was memorable for several naval duels between the frigates of the two nations. When the war broke out the United States possessed some frigates of unusual dimensions and armament. The British cruisers were quite overmatched, and in several 135instances were captured. In consequence of these disasters a new and improved class of frigate was introduced into the Royal Navy. What had happened in the case of the frigates took place also in regard to the sloops employed as cruisers. They were completely outmatched by the American vessels of corresponding class, and many of them were taken.

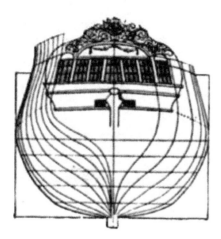

Fig. 60.—Heavy French frigate of 1780.

In 1815, on the conclusion of the long wars with France, there was, of course, a marked diminution in the number of ships built for purposes of war. The *Howe*, of 120 guns (Fig. 61), is given as an illustration of a first-rate of this period.

During the earlier years of the present century great improvements were introduced by Sir Robert Seppings and others into the structural arrangements of ships. During the long wars abundant experience had been gained as to the particular kinds of weakness which ships exhibited when exposed to the strains produced by waves. It had been felt for many years that the system of building was very defective, and the life of a man-of-war was consequently short, only fifteen years for a ship built of English oak in the Royal dockyards, and about twelve years for similar vessels built in private yards. Amongst the greatest defects was the absence of longitudinal strength to enable a ship to resist the effects of hogging and sagging strains in a sea-way.

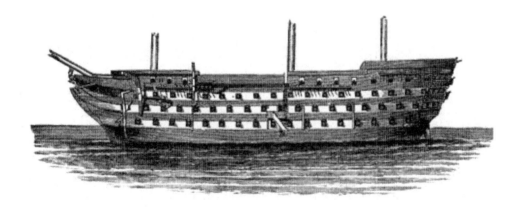

Fig. 61.—The *Howe*. 1815.

When a ship at sea is so placed that the crest of a large wave 137is passing about the midship section, the two ends may happen to be in the hollows between the waves, and in this case are to a great extent unsupported by the water, and consequently have a tendency to droop. The result is that the ship tends to arch up in the centre like a hog's back, and the upper decks are put into a state of tension, while the bottom of the vessel, on the contrary, undergoes compression. The strains set up in this way are called hogging strains. When the position of the waves is exactly reversed so that the two ends are supported by the crests, while the hollow between them passes under the middle, the latter part of the ship has a tendency to droop or sag, and the bottom is consequently extended, while the upper works are put into a state of compression.

It will be noticed, on referring to the illustration of the *Royal George* (Fig. 55), that the framework of ships built on the old system consisted of a series of transverse ribs which were connected together in the longitudinal direction by the outside planking and by the ceiling. As there was no filling between the ribs, the latter tended alternately to come closer together, or recede further apart, according as they experienced the influence of hogging or sagging stresses. The French during the eighteenth century had at various times proposed methods of overcoming this defect. One was to cross the ceiling with oblique iron riders. Another was to lay the ceiling itself and the outside planking diagonally. Sometimes the holds were strengthened with vertical and sometimes with diagonal riders, but none of these plans gave lasting satisfaction.

The means adopted by Sir Robert Seppings were as follows:—

Firstly, the spaces between the frames were filled in solid with timber (Fig. 62). In this way the bottom of the ship was transformed into a solid mass of timber admirably adapted to resist working. At the same time the customary interior planking below the orlop beams was omitted.

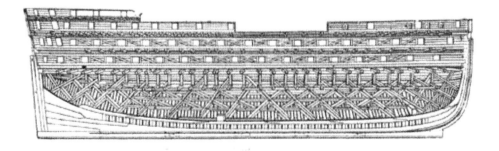

Fig. 62.—Sir Robert Seppings' system of construction.

Secondly, the beams were connected with the sides of the ship by means of thick longitudinal timbers below the knees running fore and aft, called shelf-pieces, *a, a* (Fig. 63), and similar pieces above the beams, *b, b* (Fig. 63), called waterways. These not only added to the longitudinal strength of the ship, but formed also very convenient features in the connection between the deck-beams and the ship's sides.

95

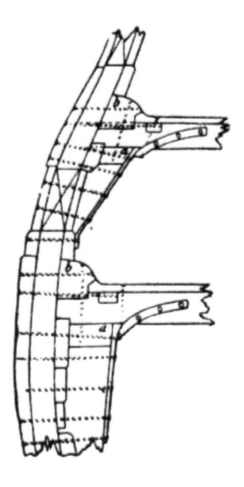

Fig. 63.—Sir Robert Seppings' system of construction.

Thirdly, a trussed frame was laid on the inside of the transverse frames in the hold of the ship. This frame consisted of diagonal riders making an angle of about 45° with the vertical, together with trusses crossing them, and longitudinal pieces, as shown in Fig. 62. This trussed frame was firmly bolted through the transverse frames and the planking of the ship.

Fourthly, it was proposed to lay the decks diagonally; but this system does not appear to have ever come into general use.

It should here be mentioned that the use of shelf-pieces and thick waterways in connection with the ends of the beams was first adopted by the French in very small vessels; also the 140 system of fillings between the frames was an extension of a method which had been in use for some time, for it was customary to fill in the spaces as far as the heads of the floors, in order to strengthen the ship's bottom against the shocks and strains due to grounding.

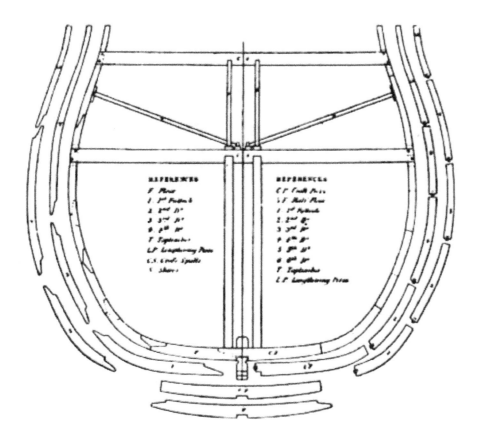

Fig. 64.—Sir Robert Seppings' system of construction.

Sir Robert Seppings further introduced many minor improvements into the details of the construction and the forms of ships. Amongst these may be mentioned the method of combining the frame-timbers. The old method of shaping the heads and heels of these timbers and of combining them with triangular chocks is shown on the left-hand side of Fig. 64. In the new method the heads and heels were cut square, and combined with circular coaks, as shown on the right-hand side in the same Fig.

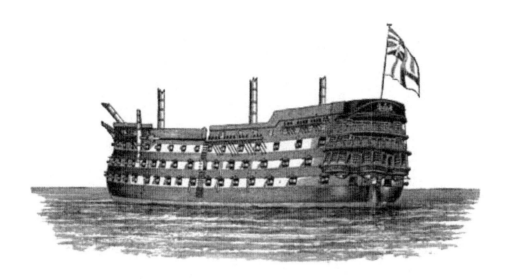

Fig. 65.—The *Waterloo*.

The principal alterations in the forms of ships introduced by Sir Robert Seppings, were connected with the shapes of the bow and stern. Hitherto the bow was cut straight across at the cathead, so as to form a vertical wall extending down to the level of the upper deck portsills, and formed of thin boarding and stanchions. The old shape of the bow is clearly shown in Figs. 52 and 55. The disadvantage of this arrangement was that it exposed the ship to the raking fire of an enemy. The old form of bow was also deficient in structural strength, and was liable to cause leakage. Sir Robert Seppings carried the rounding of the bow right up to the upper deck, and made it as strong as any other part of the ship to resist either shot or stresses. This alteration also enabled him to provide for firing several guns in a line with the keel. The old square stern was also abolished and a circular one introduced, which enabled a more powerful battery to be carried aft.

In order to bring up the account of British sailing line-of-battle ships to the period when they were superseded by the adoption of steam-power in the Royal Navy, we give illustrations of a first-rate launched in the reign of William IV., called the *Waterloo* (Fig. 65), of 120 guns, and of the *Queen* (Fig. 66), of 110 guns: the latter was the first three-decker launched in the reign of Queen Victoria. A comparison of these illustrations with those representing the largest men-of-war in the time of the Stuart sovereigns, will do more than any verbal description to show the great alterations in form and size which had taken place during two centuries. The *Waterloo* had a length on deck of 205 ft. 6 in., extreme breadth of 54 ft. 9 in., and a tonnage of 2,718; while the corresponding dimensions of the *Queen* were 204 ft. 2½ in., 55 ft. 2½ in., and 3,104 tons.

Fig. 66.—The *Queen.*

Fig. 67.—The *Thames*. East Indiaman. 1819.

During the epoch covered in this chapter the chronicles of the British Mercantile Marine were extremely meagre. The seaborne commerce of the country had increased enormously since the time of the Restoration. It had, in fact, kept pace with the development of the Royal Navy, and, in proportion145 as the naval power of the country was increased so was her commerce extended and her Mercantile Marine increased. In the year 1801 the total amount of British Mercantile shipping was about 1,726,000 tons; in 1811 it had increased to 2,163,094 tons, and in 1816 to 2,489,068; while in 1846 it had reached 3,220,685 tons. The East India Company was by far the largest mercantile shipowner and ship-hirer in the country. In the year 1772 the Company employed 33 ships of the aggregate burthen of

23,159 tons, builders' measurement. It was about this period that the Company commenced the construction of a larger type of vessel for their own use. These vessels afterwards became famous for their exploits, and were called East Indiamen. Fig. 67 is an illustration of one of them named the *Thames*, built in 1819, of 1,360 tons register. She carried 26 guns, and had a crew of 130 men.

Fig. 68.

East Indiamen were designed to serve simultaneously as freight-carriers, passenger-ships and men-of-war. In the latter capacity they fought many important actions and won many victories. Having had to fill so many purposes, they were naturally expensive ships both to build and work. Their crews were nearly four times as numerous as would be required for modern merchant sailing-ships of similar size.

At the close of the great wars in the early part of this century commercial pursuits naturally received a strong impetus. Great competition arose, not only between individual owners, but also between the shipowning classes in various countries. This caused considerable attention to be paid to the improvement of merchant-ships. The objects sought to be attained were greater economy in the working of vessels and increased speed combined with cargo-carrying capacity. The trade with the West Indies was not the subject of a monopoly as that with the East had been. It was consequently the subject of free competition amongst shipowners, and the natural 147result was the development of a class of vessel much better adapted to purely mercantile operations than were the ships owned or chartered by the East India Company. Fig. 68 is a late example of a West Indiaman, of the type common shortly after the commencement of the nineteenth century. The capacity for cargo of ships of this type was considerably in excess of their nominal tonnage, whereas in the case of the East Indiamen the reverse was the case. Also, the proportion of crew to tonnage was one-half of what was found necessary in the latter type of vessel. While possessing the above-named advantages, the West Indiamen were good boats for their time, both in sea-going qualities and in speed.

When the trade with the East was thrown open an impetus was given to the construction of vessels which were suitable for carrying freight to any part of the world. These boats were known as "Free Traders." An illustration of one of them is given in Fig. 69. They were generally from 350 to 700 tons register. The vessels of all the types above referred to were very short, relatively, being rarely more than four beams in length.

To the Americans belongs the credit of having effected the greatest improvements in mercantile sailing-ships. In their celebrated Baltimore clippers they increased the length to five and even six times the beam, and thus secured greater sharpness of the water-lines and improved speed in sailing. At the same time, in order to reduce the cost of working, these vessels were lightly rigged in proportion to their tonnage, and mechanical devices, such as capstans and winches, were substituted, wherever it was possible, for manual labour. The crew, including officers, of an American clipper of 1,450 tons, English measurement, numbered about forty.

The part played by the Americans in the carrying trade of the world during the period between the close of the great148 wars and the early fifties was so important that a few illustrations of the types of vessels they employed will be interesting. Fig. 70 represents an American cotton-ship, which also carried passengers on the route between New York and Havre in the year 1832. In form she was full and bluff; in fact, little more than a box with rounded ends.

Fig. 69.—Free-trade barque.

Fig. 70.—The *Bazaar*. American cotton-ship. 1832.

In 1840, when steamers had already commenced to cross the Atlantic, a much faster and better-shaped type of sailing-packet was put upon the New York-Havre route. These vessels were of from 800 to 1,000 tons. One of them, the *Sir John Franklin*, is shown in Fig. 71. They offered to 150passengers the advantages of a quick passage, excellent sea=going qualities, and, compared with the cotton-ships, most comfortable quarters. The Americans had also about this time admirable sailing-packets trading with British ports.

In the early fifties the doom of the sailing-packet on comparatively short voyages, such as that between New York and Western European ports, had been already sealed; but, for distant countries, such as China and Australia, and for cargo-carrying purposes in many trades, the sailing-ship was still able to hold its own. Fig. 72 represents an American three-masted clipper called the *Ocean Herald*, built in the year 1855. She was 245 ft. long, 45 ft. in beam, and of 2,135 tons. Her ratio of length to breadth was 5.45 to 1.

Fig. 73 is an illustration of the *Great Republic*, which was one of the finest of the American clippers owned by Messrs. A. Law and Co., of New York. She was 305 ft. long, 53 ft. beam, 30 ft. depth of hold, and of 3,400 tons. She was the first vessel fitted with double topsails. Her spread of canvas, without counting stay-sails, amounted to about 4,500 square yards. She had four decks, and her timber structure was strengthened from end to end with a diagonal lattice-work of iron.

Fig. 71.—The *Sir John Franklin*. American Transatlantic sailing-packet. 1840.

Fig. 72.—The *Ocean Herald*. American clipper. 1855.

The speed attained by some of these vessels was most remarkable. In 1851 the *Nightingale*, built at Portsmouth, New Hampshire, in a race from Shanghai to Deal, on one occasion ran 336 knots in twenty-four hours. In the same year the *Flying Cloud*, one of Donald McKay's American clippers, ran 427 knots in twenty-four hours in a voyage from New

York to San Francisco. This performance was eclipsed by that of another vessel belonging to the same owner, the *Sovereign of the Seas*, which on one occasion averaged over eighteen miles an hour for twenty-four consecutive hours. This vessel had a length of keel of 245 ft., 44 ft. 6 in. 153beam, and 25 ft. 6 in. depth of hold. She was of 2,421 tons register.

English shipowners were very slow to adopt these improvements, and it was not till the year 1850, after the abolition of the navigation laws, that our countrymen really bestirred themselves to produce sailing-ships which should rival and even surpass those of the Americans. The legislation in question so affected the prospects of British shipping, that nothing but the closest attention to the qualities of vessels and to economy in their navigation could save our carrying trade from the effects of American competition. Mr. Richard Green, of the Blackwall Line, was the first English shipbuilder to take up the American challenge. In the year 1850 he laid down the clipper ship the *Challenger*. About the same time, Messrs. Jardine, Matheson, and Co. gave an order to an Aberdeen firm of shipbuilders, Messrs. Hall and Co., to build two sharp ships on the American model, but of stronger construction. These vessels were named the *Stornoway* and *Chrysolite*, and were the first of the celebrated class of Aberdeen clippers. They were, however, only about half the dimensions of the larger American ships, and were, naturally, no match for them in sailing powers. The *Cairngorm*, built by the same firm, was the first vessel which equalled the Americans in speed, and, being of a stronger build, delivered her cargo in better condition, and consequently was preferred. In 1856 the *Lord of the Isles*, built by Messrs. Scott, of Greenock, beat two of the fastest American clippers in a race to this country from China, and from that time forward British merchant vessels gradually regained their ascendency in a trade which our transatlantic competitors had almost made their own.

Fig 73.—The *Great Republic*. American clipper. 1853.

It was not, however, by wooden sailing-ships that the carrying trade of Great Britain was destined to eclipse that of all her rivals. During a portion of the period covered 155in this chapter, two revolutions—one in the means of propulsion, and the other in the materials of construction of vessels—were slowly making their influence felt. About twelve years before the close of the eighteenth century the first really practical experiment was made on Dalswinton Loch, by Messrs. Miller and Symington, on the utilization of steam as a means of propulsion for vessels. An account of these experiments, and of the subsequent application and development of the invention, are given in the "Handbook on Marine Engines and Boilers," and need not, therefore, be here referred to at greater length.

The other great revolution was the introduction of iron instead of wood as the material for constructing ships. The history of that achievement forms part of the subject-matter of Part II. During the first half of the nineteenth century, good English oak had been becoming scarcer and more expensive. Shortly after the Restoration the price paid for native-grown oak was about £2 15s. a load, this being double its value in the reign of James I. The great consumption at the end of the eighteenth and the beginning of the last century had so diminished the supply, that in 1815, the year in which the great Napoleonic wars terminated, the price had risen to £7 7s. a load, which was, probably, the highest figure ever reached. In 1833 it sank to £6, and then continued to rise till, in 1850, it had reached £6 18s. per load. In consequence of the scarcity of English oak many foreign timbers, such as Dantzic and Italian oak, Italian larch, fir, pitch pine, teak, and African timbers were tried with varying success. In America timber was abundant and cheap, and this was one of the causes which led to the extraordinary development of American shipping in the first half of the nineteenth century, and it is probable that, but for the introduction of iron, which was produced abundantly and cheaply in this country, the carrying trade of the world156 would have passed definitely into the hands of the people of the United States.

The use of iron and steel as the materials for construction have enabled sailing ships to be built in modern times of dimensions which could not have been thought of in the olden days. These large vessels are chiefly employed in carrying wheat and nitrate of soda from the west coast of South America. Their structural arrangements do not differ greatly from those of iron and steel steamers which are described in Part II.

APPENDIX.

Description of a Greek Bireme of about 800 b.c.

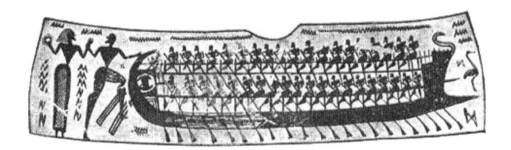

Fig. 74.—Archaic Greek bireme. About 800 b.c.

During the year 1899 the British Museum acquired a new vase of the Dipylon class, which was found near Thebes in Bœotia, and dates from about 800 b.c. On one side of the vase are represented chariots and horses, apparently about to start for a race. On the other side is a painting of a complete bireme, which, on account of its antiquity and the peculiarities of its structure is of extraordinary interest. The galley in question, Fig. 74, is reproduced from an illustration, traced direct from the vase, and published in the "Journal of Hellenic Studies," vol. xix. (1899). The chief peculiarity of the construction is that the rowers are seated upon a two-storied open staging, erected upon a very shallow hull and extending from an elevated forecastle to an equally raised structure at the stern. The stage, or platform, on which the lower tier of oarsmen is seated, is supported by vertical struts rising out of the body of the boat. The platform for the upper stage is also supported by vertical struts, which rise, not from the boat itself, but from an intermediate stage, situated between the two tiers of rowers. In the absence of a plan it is not possible to say if these platforms were floored decks, with openings cut in them, where necessary, for the legs of the rowers; or if they were simply composed of longitudinal beams connected by cross-pieces which served as seats, or benches. The latter arrangement appears to be the more probable. There are twenty oarsmen a-side, on the lower tier, and, apparently, nineteen on the upper. No attempt is made by the artist to show more than the rowers on one side, and, to avoid confusion, those on the two tiers have their oars on the opposite sides of the galley, and only one of the blades of the far side is shown. The men of the lower tier rest their feet against supports fixed to the vertical struts which support their platform, while those of the upper tier rest theirs, apparently, upon the intermediate stage. The vessel is provided with a large and a small ram, and is steered by means of two large paddles. The prow ornament resembles a snake. In some of its features, notably in the shape of the ram, the shallowness 159of the hull, and the height and number of the stages, this galley resembles the Phœnician boat of a somewhat later date, described on page 28. The arrangement of the rowers is, however, totally different in the two cases, those in the Phœnician vessel being all housed in the hull proper, while those in the Greek galley are all placed on the stages. It is a curious coincidence that the two specimens of galleys of the eighth and seventh centuries b.c., of which we possess illustrations, should both be provided with these lofty open stages.

This Greek bireme, with its shallow hull and lofty, open superstructure, could hardy have been a seaworthy vessel. The question arises, What purpose could it have been intended to serve? The rams, of course, suggest war; but the use of rams appears to have been pretty general, even in small Greek rowing-boats, and has survived into our own day in the Venetian gondola. The late Dr. A. S. Murray, keeper of the Greek and Roman antiquities at the British Museum, who wrote an account of the vase in the "Journal of Hellenic Studies," is of opinion that both the subjects on this vase represent processions, or races, held at the funeral ceremonies of some prominent citizen, and that, in fact, all the subjects on Dipylon vases seem to refer to deceased persons. He points out that Virgil mentions in the *Æneid* that games, held in honour of the deceased, commenced with a race of ships, and that he could hardly have done this if there were no authority for the practice. The large figures at the stern seem to point to the bireme of Fig. 74 being about to be used for racing purposes. The man who is going to step on board is in the act of taking leave of a woman, who holds away from him a crown, or prize, for which he may be about to contend. If this view be correct we have, at once, an explanation of the very peculiar structure of this bireme, which, with its open sides and small freeboard, could only have been intended for use in smooth water and, possibly, for racing purposes.

There are several other representations of Greek galleys, or of fragments of them, in existence. Nearly all have been found on eighth-century Dipylon vases, but, hitherto, no other specimen has been found in which all the rowers are seated on an open stage. In the collection of Dr. Sturge there is a vase of this period, ornamented with a painting of a bireme, which is as rakish and elegant in appearance as Fig. 74 is clumsy. It also is propelled by 78, or perhaps 80, rowers. Those of the lower tier are seated in the body of the boat, while those of the upper bank on what appears to be a flying deck connecting the forecastle and poop, and about 3 ft. to 3 ft. 6 in. above the seats of the lower tier.

In the Museum of the Acropolis there are also some fragments of Dipylon vases, on which are clearly visible portions of biremes. The rowers of the lower bank are here again, seated in the hull of the galley160 and appear to be working their oars in large square portholes, while the upper row are seated on a flying deck, the space between which and the gunwale of the hull is partly closed in by what appear to be patches of awning or light fencing. The portholes above referred to are in fact merely open intervals between the closed-in spaces. Similar lengths of fencing may be seen in the representation of a Phœnician galley (Fig. 7, p. 27).

From the above description it is not difficult to see how the galley, with two tiers of oars, came to be evolved from the more primitive unireme. First, a flying deck was added for the accommodation of the upper tier of rowers. It formed no part of the structure of the ship, but was supported on the latter by means of struts, or pillars. The spaces between the hull and the flying deck at the two ends of the galley were closed in by a raised forecastle and poop. These additions were necessary in order to keep the vessel dry, and attempts were no doubt made to give protection to the remainder of the sides by means of the patches of light awning mentioned above. The step from this to carrying the structure of the sides up bodily, till they met the upper deck, and of cutting portholes for the lower tier of oars, would not be a long one, and would produce the type of bireme illustrated on p. 31 (Fig. 9).161

FOOTNOTES:

[1] This illustration is taken from Mr. Villiers Stuart's work, "Nile Gleanings."

[2] "A History of Egypt under the Pharaohs," by Dr. Henry Brugsch Bey. Translated and edited from the German by Philip Smith, B.A.

[3] "Nile Gleanings," p. 309.

[4] The inscription is taken from the "History of Egypt under the Pharaohs," by Dr. Henry Brugsch Bey. Translated and edited by Philip Smith, B.A. Second edition, pp. 137, 138.

[5] "A History of Egypt under the Pharaohs," by Dr. Henry Brugsch Bey. Translated and edited from the German by Philip Smith, B.A. Second edition, p. 358.

[6] Egypt Exploration Fund: *Archæological Report*, 1895-1896. Edited by F. L. Griffith, M.A.

[7] "The History of Herodotus," translated by G. C. Macaulay, M.A. 1890. Vol. i. p. 157. (ii. 96 is the reference to the Greek text.)

[8] In Appendix, p. 157, will be found an account of an eighth-century Greek bireme, recently discovered.

[9] For latest information on Greek vessels of Archaic period, *see* Appendix.

[10] This figure is obtained by adding the height of the lowest oar-port above the water, viz. 3 ft., to 2 ft. 6 in., which is twice the minimum vertical interval between successive banks.

[11] This illustration is taken from Charnock's "History of Marine Architecture." It is copied by Charnock from Basius, who, in his turn, has evidently founded it on the sculptures on Trajan's Column.

[12] "Cæsar, de Bello Gallico," bk. iii. chap. 13.

[13] Vol. xxii., p. 298. Paper by Mr. Colin Archer.

[14] "Archéologie Navale."

[15] W. S. Lindsay, "History of Merchant Shipping and Ancient Commerce," vol. ii. p. 4.

[16] The details, as related by various authorities, differ slightly.

[17] According to some accounts there were 1,497 bronze and 934 iron guns of all calibres.

Made in the USA
Las Vegas, NV
14 December 2021

37679951R00063